A PLANET OF VIRUSES

A PLANET OF
VIRUSES

Third Edition

CARL ZIMMER

ILLUSTRATIONS BY IAN SCHOENHERR

THE UNIVERSITY OF CHICAGO PRESS
CHICAGO AND LONDON

CARL ZIMMER is a columnist for the *New York Times*, writes for the *Atlantic* and other magazines, and is the author of 14 books, including *Life's Edge* and *She Has Her Mother's Laugh*. He is professor adjunct in the Department of Molecular Biophysics and Biochemistry at Yale University, where he teaches writing about science.

The University of Chicago Press, Chicago 60637
The University of Chicago Press, Ltd., London

First edition published 2011. Second edition 2015. Third edition 2021.
Printed in the United States of America

30 29 28 27 26 25 24 23 22 21   1 2 3 4 5

ISBN-13: 978-0-226-78259-1 (paper)
ISBN-13: 978-0-226-78262-1 (e-book)
DOI: https://doi.org/10.7208/chicago/9780226782621.0001

Some of the essays in this book were written for the World of Viruses project, funded by the National Center for Research Resources at the National Institutes of Health through the Science Education Partnership Award (SEPA) Grant No. R25 RR024267 (2007–2012). Its content is solely the responsibility of the authors and does not necessarily represent the official views of NCRR or NIH. Visit http://www.worldofviruses.unl.edu for more information and free educational materials about viruses. World of Viruses is a project of the University of Nebraska–Lincoln.

Library of Congress Cataloging-in-Publication Data
Names: Rhodes, John, 1947– author.
Title: How to make a vaccine : an essential guide for COVID-19 and beyond / John Rhodes.
Description: Chicago : University of Chicago Press, 2021. | Includes bibliographical references and index.
Identifiers: LCCN 2020053338 | ISBN 9780226792514 (paperback) | ISBN 9780226792651 (ebook)
Subjects: LCSH: COVID-19 (Disease) | Vaccines.
Classification: LCC RA644.C67 R48 2021 | DDC 614.5/92414—dc23
LC record available at https://lccn.loc.gov/2020053338

♾ This paper meets the requirements of ANSI/NISO Z39.48-1992 (Permanence of Paper).

TO GRACE,
my favorite host

CONTENTS

FOREWORD

Viruses wreak chaos on human welfare, affecting the lives of almost a billion people. They have also played major roles in the remarkable biological advances of the past century. The smallpox virus was humanity's greatest killer, and yet it is now one of the few diseases to have been eradicated from the globe. Emerging and re-emerging viruses, such as those that cause influenza, Ebola, Zika, and now the global COVID-19 pandemic, pose global catastrophic threats and extraordinary challenges. These and other viruses will likely continue to threaten human well-being. A better understanding of these viruses will help us to prepare and prevent future viral diseases and pandemics.

Viruses are unseen but dynamic players in the ecology of Earth. They move DNA between species, provide new genetic material for evolution, and regulate vast populations of organisms. Every species, from tiny microbes to large mammals, is influenced by the actions of viruses. Viruses extend their impact beyond species to affect climate, soil, the oceans, and fresh water. When you consider how every animal, plant, and microbe has been shaped through the course of evolution, one has to consider the influential role played by the tiny and powerful viruses that share this planet.

After the first edition of *A Planet of Viruses* was published in

2011, viruses continued to surprise us all. The Ebola virus, once limited to small flare-ups in remote parts of Africa, exploded into massive outbreaks and, for the first time, spread to other continents. New viruses, like MERS and SARS, leaped from animals to humans through zoonotic infections. HIV, first recognized in 1983, has now infected almost 38 million people throughout the world. But scientists are also discovering new ways to harness the amazing diversity of viruses for our own benefit. Carl Zimmer has drawn on all these developments to produce this new edition of *A Planet of Viruses*.

Zimmer originally wrote most of these essays for the World of Viruses project as part of a Science Education Partnership Award (SEPA) at the National Institutes of Health (NIH). World of Viruses was created to help people understand more about viruses and virology research through comics, teacher professional development, mobile phone and iPad applications, and other materials. For more information about World of Viruses, visit http://worldofviruses.unl.edu.

JUDY DIAMOND, PHD
Professor and Curator, University of Nebraska State Museum
Director, World of Viruses Project

CHARLES WOOD, PHD
Lewis L. Lehr University Professor of Biological Sciences and Biochemistry, University of Nebraska
Director, Nebraska Center for Virology

INTRODUCTION

"A Contagious Living Fluid"

TOBACCO MOSAIC VIRUS AND THE DISCOVERY OF THE VIROSPHERE

Fifty miles southeast of the Mexican city of Chihuahua is a dry, bare mountain range called Sierra de Naica. In 2000, miners worked their way down through a network of caves below the mountains. When they got 1,000 feet underground, they found themselves in a place that seemed to belong to another world. They were standing in a chamber 30 feet wide and 90 feet long. The ceiling, walls, and floor were lined with smooth-faced, translucent crystals of gypsum. Many caves contain crystals, but not like the ones in Sierra de Naica. They measured up to 36 feet long and weighed as much as 55 tons. These were not crystals to hang from a necklace. These were crystals to climb like hills.

Since its discovery, a few scientists have been granted permission to visit this extraordinary chamber, known now as the Cave of Crystals. Juan Manuel García-Ruiz, a geologist at the University of Granada, was one of them. After studying the crystals, he determined that they formed 26 million years ago. At the time, molten rock was rising up from deep inside the Earth, building the mountains. Subterranean chambers took shape and filled with hot, mineral-laced water. The heat of the underlying magma kept the water at a scalding 136°F. That was the ideal temperature for the minerals to settle out of the water and form crystals. For reasons that aren't clear, the water stayed at that perfect temperature for hundreds of thousands of years. That long simmer allowed the crystals to grow to surreal sizes.

In 2009, another scientist, Curtis Suttle, led a new expedition to the Cave of Crystals. Suttle and his colleagues scooped up water from the chamber's pools and brought it back to their laboratory at the University of British Columbia to analyze. When you consider Suttle's line of work, his journey might seem like a fool's errand. Suttle has no professional interest in crystals, or minerals, or any rocks at all for that matter. He studies viruses.

There are no people in the Cave of Crystals for the viruses to infect. There are not even any fish. The cave has been effectively cut off from the biology of the outside world for millions of years. Yet Suttle's trip was well worth the effort. After he prepared his samples of crystal water, he gazed at them under a microscope. He saw viruses—swarms of them. There are as many as 200 million viruses in every drop of water from the Cave of Crystals.

That same year another scientist, Dana Willner, led a virus-hunting expedition of her own. Instead of a cave, she dove into the human body. Willner had people cough up sputum into a cup, and out of that fluid she and her colleagues fished out fragments of DNA. They compared the DNA fragments to millions of sequences stored in online databases. Much of the DNA was human, but many fragments came from viruses. Before Willner's expedition, scientists had assumed the lungs of healthy people were sterile. Yet Willner discovered that, on average, people have 174 species of viruses in their lungs. Only 10 percent of the species Willner found bore any close kinship to any virus ever found before. The other 90 percent were as strange as anything lurking in the Cave of Crystals.

In caves and in lungs, in glaciers in Tibet and in winds flowing high over mountains, scientists keep discovering viruses. They are finding them faster than they can make sense of them. So far, scientists have officially named a few thousand species of viruses, but the true total may, by some estimates, reach into the trillions. Virology is a science in its infancy. Yet viruses themselves are old companions. For thousands of years, we knew viruses only from their effects in sickness and death. Until recently, however, we did not know how to join those effects to their cause.

The very word *virus* began as a contradiction. We inherited the word from the Roman Empire, where it meant, at once, the venom of a snake or the semen of a man. Creation and destruction in one word.

Over the centuries, *virus* took on another meaning: it signified any contagious substance that could spread disease. It might be a fluid, like the discharge from a sore. It might be a substance that traveled mysteriously through the air. It might even impregnate a piece of paper, spreading disease with the touch of a finger.

Virus began to take on its modern meaning only in the late 1800s, thanks to an agricultural catastrophe. In the Netherlands, tobacco farms were swept by a disease that left plants stunted, their leaves a mosaic of dead and live patches of tissue. Entire farms had to be abandoned.

In 1879, Dutch farmers came to a young agricultural chemist named Adolph Mayer to beg for help. Mayer gave the scourge a name: tobacco mosaic disease. To find its cause, he investigated the environment in which the plants grew—the soil, the temperature, the sunlight—but he could find nothing to distinguish the healthy plants from the sick ones. Perhaps, Mayer thought, an invisible infection was to blame. Scientists had already discovered that fungi could infect potatoes and other plants, so Mayer looked for fungus on the tobacco plants. He found none. He looked for parasitic worms infesting the leaves. Nothing.

Finally Mayer extracted the sap from sick plants and injected drops into healthy tobacco. The healthy plants turned sick. Mayer realized that some microscopic pathogen must be multiplying inside the tobacco. He took sap from sick plants and incubated it in his laboratory. Colonies of bacteria began to grow. They became large enough that Mayer could see them

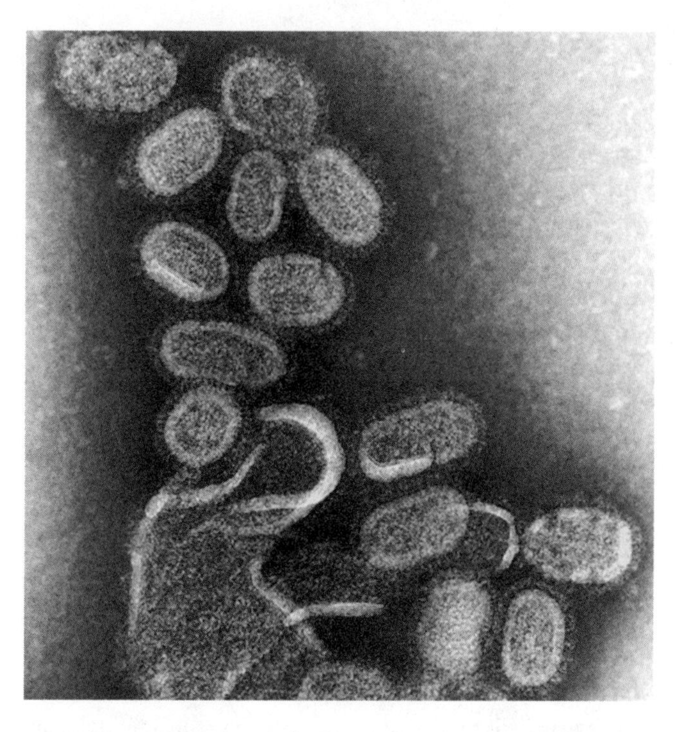

Tobacco mosaic viruses, which cause plant diseases worldwide.

with his naked eye. Mayer applied these bacteria to healthy plants, wondering if they would trigger tobacco mosaic disease. They did nothing of the sort. Scientists have since learned that plants are coated in bacteria from their leaves down to their roots. Far from making plants sick, many of the microbes help them thrive. And with that failure, Mayer's research ground to a halt. The world of viruses remained unopened.

A few years later, another Dutch scientist named Martinus Beijerinck picked up where Mayer left off. He wondered if something other than worms, fungi, or bacteria was responsible for tobacco mosaic disease—something far smaller. He ground up diseased plants and passed the substance through a filter so fine that it blocked all the cells it contained. He was left with a clear, cell-free fluid. When Beijerinck injected it into healthy plants, they developed tobacco mosaic disease. And when Beijerinck filtered the juice from newly infected tobacco leaves, he could pass the disease on to more healthy plants.

In 1898, Beijerinck describe his findings, referring to his filtered juice as a "contagious living fluid." It carried something inside it that spread tobacco mosaic disease. That substance was living, Beijerinck assumed, but it had to be different from life as he knew it. In the late 1800s, researchers were confident that all living things are made of cells. His fluid contained none. And whatever it did contain must be remarkably durable. Beijerinck could add alcohol to the filtered fluid, and it would remain infective. Heating the fluid to near boiling did it no harm. Beijerinck soaked filter paper in the infectious sap and let it dry. Three months later, he could dip the paper in water and use the solution to sicken new plants.

Beijerinck gave a name to the mysterious agent in his contagious living fluid: *virus*. He borrowed this ancient word and gave it a new meaning. Yet Beijerinck could not define what a virus actually was. All he could do was say what it was *not*. It was not an animal, a plant, a fungus, or a bacterium. It was something else.

It soon became clear that Beijerinck had discovered just one kind of virus. In the early 1900s, other scientists used his method of filtering and infecting to discover other viruses that caused other diseases. Eventually, they learned how to cultivate some viruses outside of their hosts. If they could grow colonies of cells in a Petri dish, they could grow viruses, too.

Even then, scientists *still* couldn't agree about what viruses actually were. Some argued viruses were parasites that exploited cells. Some maintained they were just chemicals. The confusion over viruses was so profound that scientists could not even agree if viruses were living or dead. In 1923, the British virologist Frederick Twort declared, "It is impossible to define their nature."

That confusion began to disperse with the work of a chemist named Wendell Stanley. As a chemistry student in the 1920s, Stanley learned how to make crystals by combining molecules into repeating patterns. Scientists could learn things about molecules in their crystal form that they could not otherwise. They fired X-rays at the crystals, which bounced off the atoms and then struck photographic plates. The X-rays left behind repeating patterns of curves, lines, and dots, which scientists could then use to determine the structure of the molecules in the crystal.

In the early 1900s, crystals helped solved one of biology's biggest mysteries. At the time, scientists knew that living things contained puzzling molecules called enzymes, which could precisely break down certain other molecules. To figure out the true nature of enzymes, scientists turned them into crystals. The signature of their X-rays revealed that they were made of proteins. Thinking of the transforming power of viruses, Stanley wondered if they were made of proteins, too.

To find out, he turned viruses into crystals. He chose a familiar species for his work: the tobacco mosaic virus. Stanley collected the juice of infected tobacco plants and then passed it through fine filters, as Beijerinck had done four decades earlier. He removed every bit of contamination from the fluid and then prepared it to crystalize. To his amazement, tiny needles began to form in the fluid. They grew into opalescent sheets. For the first time in history, viruses had become visible to the naked eye.

Stanley found that his virus crystals were rugged as a mineral. He could store them away for months like table salt in a pantry. When he then added the crystals to water, they vanished from view, becoming a contagious living fluid once more.

Stanley's experiment, which he published in 1935, dazzled the world. "The old distinction between death and life loses some of its validity," declared the *New York Times*.

But Stanley's work, while groundbreaking, had its limits. For one thing, he made a small but profound mistake. Tobacco mosaic viruses were not made of pure protein. The British scientists Norman Pirie and Fred Bawden discovered in 1936 that 5 percent of the virus's weight was made up of another molecule, a mysterious strand-shaped substance called nucleic acid. Nucleic acids, scientists would later discover, are the stuff of genes, the instructions for building proteins and other molecules. Our cells store their genes in double-stranded nucleic acids, known as deoxyribonucleic acid, or DNA for short. Many viruses have DNA-based genes as well. Other viruses, such as tobacco mosaic virus, have a single-stranded form of nucleic acids, called ribonucleic acid, or RNA. It would take decades for scientists to discover how viruses used this genetic material to take over cells and get them to make new viruses.

And while Stanley saw viruses for the first time, he only saw them in droves. Each crystal he created might contain millions of tobacco mosaic viruses nestled together in an interlocking grid. To see individual viruses, scientists first had to invent a new generation of microscopes, ones that use a beam of electrons to reveal tiny objects. In 1939, Gustav Kausche, Edgar Pfannkuch, and Helmut Ruska mixed tobacco mosaic crystals into drops of purified water and put them under one of the new devices. They could see minuscule rods, each measuring about 300 nanometers long.

No one had ever seen a living thing anywhere near so small. To contemplate the size of viruses, tap out a single grain of salt onto a table. Stare at the tiny cube. You could line up about 10 skin cells along one side of it. You could line up about 100 bacteria. And you could line up 1,000 tobacco mosaic viruses, end to end, alongside that same grain of salt.

In the decades that followed, virologists went on to dissect viruses, to map their molecular geography. While viruses contain nucleic acids and proteins like our own cells, they use these molecules in a profoundly different way. A human cell is stuffed with millions of different molecules that are in constant quivering motion, cleaving each other apart or bonding each other together, which the cell uses to sense its surroundings, crawl, take in food, grow, and decide whether to divide in two or kill itself for the good of its fellow cells. Virologists found that viruses, as a rule, were far simpler. They typically were just protein shells holding a few genes. Virologists discovered that viruses can replicate themselves, despite their paltry genetic instructions, by hijacking other forms of life. They inject their genes and proteins into a host cell, which they manipulate into producing new copies of themselves. One virus goes into a cell, and within a day thousands of viruses may come out.

By the 1950s, virologists had grasped these fundamental facts. But that understanding did not bring virology to a halt. For one thing, virologists knew little about the many different ways in which viruses make us sick. They didn't know why papillomaviruses can cause horns to grow on rabbits and cause hundreds of thousands of cases of cervical cancer each year. They didn't know what made some viruses deadly and others relatively harmless. They had yet to learn how viruses evade the defenses of their hosts and how they evolve faster than anything else on the planet. In the 1950s they did not know that a virus had spread decades earlier from chimpanzees and other primates into humans, a virus that would go on to become one of the greatest killers in history, known as HIV. Nor could they predict that in 2020 a new virus called SARS-CoV-2 would sweep the planet, plunging the global economy into its worst crisis since the Great Depression.

In the 1950s, scientists also didn't know how important viruses are beyond disease. They could not have dreamed of the vast number of viruses that exist on Earth; they could not have guessed that much of life's genetic diversity is carried in viruses. They did not know that viruses help produce much of the oxygen we breathe and help control the planet's thermostat. And they certainly would not have guessed that the human genome is partly composed from thousands of viruses that infected our distant ancestors, or that life as we know it may have gotten its start four billion years ago from viruses.

Now scientists know these things—or, to be more precise, they know *of* these things. They now recognize that from the Cave of Crystals to our own interiors, Earth is a planet of viruses. Their understanding is still rough, but it is a start.

So let us start as well.

OLD COMPANIONS

The Uncommon Cold

HOW RHINOVIRUSES GENTLY CONQUERED THE WORLD

Around 3,500 years ago, an Egyptian physician sat down and wrote the oldest known medical text. Among the diseases he described was something called *resh*. Even with that strange-sounding name, its symptoms—a cough and a flowing of mucus from the nose—are immediately familiar to us all. *Resh* is the common cold.

Some viruses that beset us today are new to humanity. Other viruses are obscure and exotic. But human rhinoviruses—the chief cause of the common cold—are old companions. It's been estimated that every human being on Earth will spend a year of his or her life lying in bed, sick with colds. The human rhinovirus is, in other words, one of the most successful viruses of all.

Before the discovery of rhinoviruses, doctors floundered to explain the cause of colds. Hippocrates, the ancient Greek physician, blamed an imbalance of the humors. Two thousand years later, in the early 1900s, our knowledge of colds hadn't improved much. The physiologist Leonard Hill declared that colds were caused by walking outside in the morning.

In 1914, a German microbiologist named Walther Kruse gained the first solid clue about the origin of colds by having a snuffly assistant blow his nose. Kruse mixed the assistant's mucus into a salt solution, poured it through a filter, and then put a few drops of the filtered fluid into the noses of 12 colleagues. Four of them came down with colds. Later, Kruse did the same thing to 36 students, and 15 of them got sick. While he ran this experiment Kruse also kept track of 35 people who didn't get the drops. Only one of them came down with a cold on their own. Kruse's experiments made it clear that the drops from people with colds contained a tiny pathogen that was responsible for the disease.

At first, many experts believed it was some kind of bacteria. But the American physician Alphonse Dochez ruled that out in 1927. He filtered the mucus from people with colds, just as Beijerinck had filtered tobacco plant sap 30 years before. Even with the bacteria removed, the fluid could still make

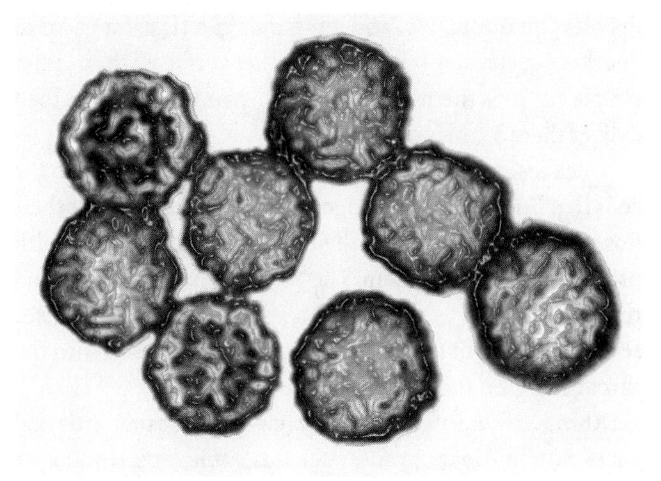

Rhinoviruses, the most common cause of colds

people sick. Only a virus could have slipped through Dochez's filters.

It took another three decades before scientists figured out exactly which viruses had slipped through. The most common of them are known as human rhinoviruses (*rhino* means nose). Rhinoviruses are remarkably simple. While we humans have about 20,000 genes, rhinoviruses have only 10. And yet this haiku of genetic information is enough to let rhinoviruses invade our bodies, outwit our immune system, and produce new viruses that can escape to new hosts.

To get to those new hosts, rhinoviruses travel in droplets. They can get into the tiny ones that we exhale with each breath. They can get into the bigger droplets we blast out when we sneeze or cough. A careless wipe of the nose can put those droplets on our hands, and our hands can transfer them to doorknobs, elevator buttons, and other surfaces where other people can pick them up on their own hands, which can then deliver them to their own noses.

Once inside a fresh nose, rhinoviruses can latch on to the cells that line the nasal passage. They slip inside and use their host cells to make copies of their genetic material, along with protein shells to hold them. The host cell then rips apart, and the new rhinoviruses escape. In some hosts, rhinoviruses remain limited to the nose, but in others they slip into the throat and even the lungs.

Rhinoviruses infect relatively few cells, causing little real harm. So why can they cause such miserable experiences? We have only ourselves to blame. Infected cells release signaling molecules, called cytokines, which attract nearby immune cells. Those immune cells then make us feel awful. They create inflammation that triggers a scratchy feeling in the throat and leads to the production of mucus around the site of the infection. In order to recover from a cold, we have to wait not only for the immune system to wipe out the virus, but also for the immune system itself to calm down.

In ancient Egypt, physicians treated *resh* by dabbing a mixture of honey, herbs, and incense around the nose. Fifteen centuries later, the Roman scholar Pliny the Elder recommended rubbing a mouse against the nose instead. In seventeenth-century Europe, some physicians used a blend of gunpowder and eggs, others a mixture of suet and fried cow dung. Leonard Hill recommended starting the day with a cold shower.

None of these treatments worked, but even today we lack a proven cure for the common cold. In the late 1900s, some researchers got encouraged by the discovery that zinc could stop rhinoviruses from infecting cells grown in Petri dishes. Before long, drug stores were selling zinc tablets without a prescription, even though no one had yet shown that they worked in actual people. Some small clinical studies later hinted that zinc might cut a cold down by a couple days. But when a Finnish scientist named Harri Hemilä led a carefully designed trial on 253 volunteers, he found no benefit. In fact, Hemilä reported in 2019, the volunteers who took zinc tablets took a little longer to recover from a cold than people who took sugar pills.

Other common treatments for the cold may not only be useless—they may even cause harm. Parents often give children cough syrup for colds, but studies show it doesn't make people get better faster. In fact, cough syrup poses a wide variety of rare yet serious side effects, such as convulsions, rapid heart rate, and even death. The US Food and Drug Administration warns that children under the age of two—who get colds the most often—should not take cough syrup.

It's also a mistake to treat a cold with antibiotics. Antibiotics are designed to kill bacteria and are useless against viruses. Doctors prescribe them depressingly often for colds anyway. In some cases it may be hard to tell from the symptoms patients display whether they're infected with rhinoviruses or bacteria. In other cases, doctors may respond to pressure from worried parents to do *something*. The harm that antibiotics cause in these cases isn't limited to one patient: we suffer. Our bodies are home to trillions of harmless bacteria, and antibiotics can foster the evolution of resistant strains. Those resistant bacteria can pass on their genes to disease-causing microbes. As a result, when we need antibiotics to work, they may fail us.

One reason the cold remains so hard to treat may be that we've underestimated the rhinovirus. It exists in many forms, and scientists are only starting to get a true reckoning of its genetic diversity. As a cell makes new rhinoviruses, it typically makes mistakes in copying the virus's genes. Over the generations, the virus's lineages become increasingly different. By the end of the twentieth century, scientists had identified dozens of strains of rhinovirus. They belonged to two great lineages, known as HRV-A and HRV-B.

In 2006, Ian Lipkin and Thomas Briese of Columbia University discovered that some New Yorkers with flu-like symptoms were infected with rhinoviruses that did not belong to either HRV-A or HRV-B. Instead, they formed a previously unknown third lineage, which Lipkin and Briese dubbed HRV-C. Since their discovery, researchers have found HRV-C all around the world.

The more strains scientists discover, the better they come to understand the evolutionary history of rhinoviruses. Some of their genes turn out to be evolving very quickly as the viruses outrace our immune systems. One type of weapon we use to fight against viruses is antibodies—molecules that can latch on to the surface of a virus and disrupt it in all sorts of ways. Mutations can alter the surface of rhinoviruses such that those antibodies can no longer stick. Our immune systems can make new antibodies, but new mutations can allow the viruses to escape once more.

This rapid evolution has helped create a tremendous diversity of rhinoviruses. Each of us can expect to get infected by several different human rhinovirus strains every year. And just as this evolution frustrates our immune systems, it also frustrates researchers who are trying to make antivirals that can cure colds. If an antiviral works well against one strain of rhinovirus, it may fail against others. And there's always a chance that a new mutation may enable a rhinovirus to resist the drug and explode in numbers while other viruses die off.

While we don't have a cure yet for the common cold, we shouldn't give up in despair. Although some parts of rhinoviruses evolve rapidly, other parts barely change at all. In these regions of a rhinovirus, mutations may be lethal. If scientists can target these vulnerable spots in the rhinovirus, they may be able to take on every rhinovirus on Earth.

But should they? The answer is actually not clear. Human rhinoviruses impose a serious burden on public health, not just by causing colds, but by opening the way for more harmful pathogens. Yet the effects of human rhinovirus itself are relatively mild. Most colds finish in under a week, and 40 percent of people who test positive for rhinoviruses suffer no symptoms at all. In fact, human rhinoviruses may offer some benefits to their human hosts. Scientists have gathered a great deal of evidence that children who get sick with relatively harmless viruses and bacteria may be protected from immune disorders when they get older, such as allergies and Crohn's disease. Human rhinoviruses may help train our immune systems not to overreact to minor triggers, instead directing their assaults to real threats. Perhaps we should not think of colds as ancient enemies but as wise old tutors.

Looking Down from the Stars

INFLUENZA'S NEVER-ENDING REINVENTION

Influenza. If you close your eyes and say the word aloud, it sounds lovely. It would make a good name for a picturesque ancient Italian village. *Influenza* is, in fact, Italian (it means *influence*). And the name is indeed ancient, dating back to the Middle Ages. But the charming associations stop there. The disease got its name from the belief among medieval physicians that stars influenced the health of their patients. They

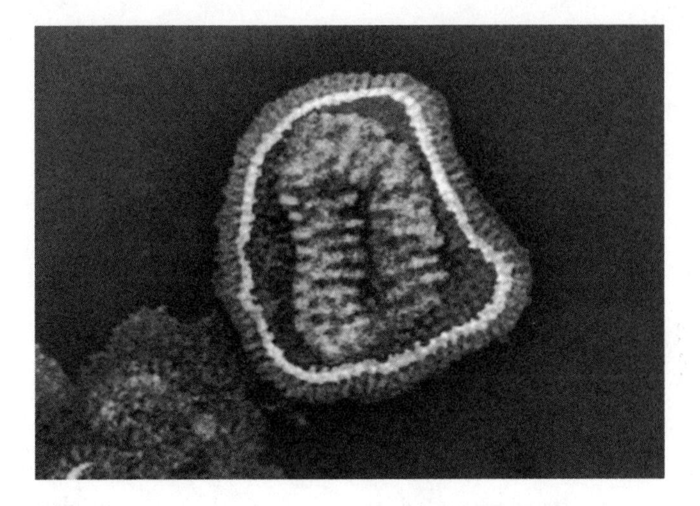

Influenza virus: the envelope layer and capsid with RNA segments inside

could trigger a debilitating fever, one that could turn into vast outbreaks every few decades.

Influenza has continued to burden the world with its devastation. In 1918, a particularly virulent outbreak of the flu spread across the planet and killed an estimated 50 to 100 million people. Even in ordinary years, influenza takes a brutal toll. The World Health Organization estimates that the flu annually strikes a billion people, killing somewhere between 290,000 and 650,000 of them.

Today scientists know that influenza is the work not of the heavens but of a microscopic virus. Like cold-causing rhinoviruses, influenza viruses manage to wreak their harm with very little genetic information—just 13 genes. They spread in the droplets sick people release with their coughs, sneezes, and runny noses. Once a flu virus gets into the nose or throat, it can latch on to a cell lining the airway and slip inside. As flu viruses spread from cell to cell, they leave destruction in their wake. The mucus and cells lining the airway get destroyed, as if the flu viruses were a lawn mower cutting grass.

For most people, this havoc lasts for only a few days. We have our immune systems to thank for that. Just as our immune system can learn to make antibodies against rhinoviruses, they can learn to make antibodies against influenza viruses, targeting their own unique proteins. One of the most common ways that antibodies protect us from the flu is by latching on to the very tip of proteins that project from the surface of the viruses. The viruses use these tips to latch on to cells and invade them. Antibodies prevent them from gaining entry, a bit like putting a dab of gum on the end of a key so that it can no longer turn a lock.

Unfortunately, an antibody that works on one kind of flu virus may not work on another. There are over 130 subtypes of influenza circulating among human beings, and each flu season a few of them dominate the viral population. If you already have antibodies to a subtype, it won't get you sick. If another subtype strikes, you may get sick for a while as you create new antibodies that can stop it. The virus can use this time to spread into the lungs, causing more damage. Normally, the top layer of cells serves as a barrier against a wide array of pathogens. The pathogens get trapped in the mucus, and the cells snag them with hairs, swiftly notifying the immune system of intruders. Once the influenza lawn mower has cut away that protective layer, pathogens can slip in and cause dangerous lung infections, some of which can be fatal.

The flu vaccine can dramatically lower the odds of such a tragic outcome. It is made of the proteins that stud the surface of influenza viruses, prompting the immune system to prepare antibodies before we get infected with real viruses. Unfortunately, the vaccine has to match a subtype of influenza to provide the strongest protection. Since subtypes churn so much from one year to the next, we have to get a flu shot at the start of every flu season to keep our defenses up to date.

To track this relentless churn, scientists collect viruses from patients around the world and read the sequence of their genes. They observe new mutations arising, creating tiny changes to influenza proteins. They also track a viral version of sex happening in our airways. When a flu virus hitches a ride aboard a droplet and infects a new host, it sometimes invades a cell that's already harboring another flu virus. And when two different flu viruses reproduce inside the same cell, things can get messy.

The genes of a flu virus are stored on eight separate segments. When a host cell starts manufacturing the segments from two different viruses at once, they sometimes get mixed together. The new offspring end up carrying genetic material from both viruses, in a mixing known as reassortment. When humans have children, the parents' genes are mixed together, creating new combinations of the same two sets of DNA. Reassortment allows flu viruses to mix genes together into new combinations of their own—combinations that can let them evade our immune system and spread more swiftly from person to person.

Every few decades, this ordinary churn of mutation and reassortment is interrupted by something far worse: a pandemic. A new influenza subtype emerges and sweeps across the world, causing a wave of death. The 1918 pandemic was the first of the twentieth century, and it was followed by others in 1957 (causing one to two million deaths), 1968 (700,000 deaths), and 2009 (363,000 deaths).

These new flu subtypes did not come to us from the stars, however. They came to us from birds. Over 100 different species of birds can be infected by influenza viruses. They carry all known strains of the human influenza virus, along with a vast diversity of other flu viruses that don't infect humans (at least not yet). Rather than infecting the birds' airways, flu viruses typically infect their guts, where they can lurk harmlessly. The viruses are shed in bird droppings and then infect healthy birds that ingest virus-laden water.

Sometimes bird flu ends up in people instead. They may pick up a virus on a chicken farm or at a poultry market. The receptors the influenza viruses used to get into cells in bird guts are similar in shape to those in our airways. Bird flu viruses can sometimes latch on to those receptors and slip inside.

Most of these invasions end in failure, though. The genes a bird flu virus needs to thrive are different from those needed inside a human body. Human bodies are cooler than bird bodies, for example, and that difference means that molecules need different shapes to run efficiently. As a result, bird flu viruses may replicate slowly in our bodies, making them easy prey for our immune systems. In addition, they are adapted to bird guts and water, making them poorly suited to spreading in droplets from one person to another. The result of this mismatch is that bird flu rarely manages to get from one person to another. Starting in 2005, for example, a strain of bird flu called H5N1 began to sicken hundreds of people in Southeast Asia. While it was very dangerous to those who got infected, it never once managed to move from one person to another.

But every now and then, bird flu viruses manage to adapt to our bodies. They may gain mutations that let them use our cells to build new viruses more quickly. They can capture entire genes through reassortment, turning into bird-human flu hybrids. A new strain produced from one of these combinations may be able to easily spread from person to person. But because it has never circulated among humans before, no one has any defenses that could slow its spread.

The origins of each flu pandemic remain murky, but the best understood of them is the most recent, in 2009. Its history reaches all the way back to the great pandemic of 1918. The subtype that emerged back then, known as H1N1, made its way from humans to pigs, and it continued infecting pigs long after the human pandemic ended. As pigs were shipped from country to country through international trade, the H1N1 subtype spread into new herds, mutating along the way.

In the 1990s, pigs from both Europe and North America were important to Mexico. Each stock of animals carried its own version of H1N1. And in Mexico, those two kinds of influenza mixed their genes through reassortment in pigs. Later, that reassorted H1N1 virus shuffled genes with another subtype called H3N2, possibly originating from birds. This three-way hybrid continued to circulate among Mexican pigs for years. Scientists estimate that it finally jumped into humans in the fall of 2008. It then spread quietly among people for months before finally coming to light the following spring.

Public health workers were terrified by the emergence of

this new flu, which they dubbed Human/Swine 2009 H1N1. It was impossible to predict in advance just what it might do. Would it prove as bad as 1918? Even if it caused just a fraction of the previous death toll, that would be a catastrophe. And so public health organizations launched a global campaign to protect people from infection.

Unfortunately, the virus proved remarkably contagious. And equally unfortunately, it took several months to make a new vaccine against 2009 H1N1, which only provided moderate protection. As a result, 2009 H1N1 managed to spread from country to country, infecting 10 to 20 percent of all people on Earth. But to the surprise and relief of scientists, it proved to be comparatively mild. While the loss of 363,000 lives cannot be ignored, it's important to recognize that the toll could have been far greater.

As I write in 2021, we await the next flu pandemic. Influenza viruses are mixing and evolving in billions of birds—on turkey farms, on beaches, on migration stopovers across the world. One day they will wind up in a new recipe. Whether it will be mild like 2009 or disastrous like 1918, scientists cannot predict. But we are not helpless as we wait to see what evolution has in store for us. All of us can do things to slow the spread of the flu, such as washing our hands. And scientists are learning how to make more effective vaccines by tracking the evolution of the flu virus so they can do a better job of predicting which strains will be most dangerous in flu seasons to come. We may not have the upper hand over the flu yet, but at least we no longer have to look to the stars to defend ourselves.

Rabbits with Horns

HUMAN PAPILLOMAVIRUS AND INFECTIOUS CANCER

Stories about rabbits with horns circulated for centuries. Eventually they crystallized into the myth of the jackalope. If you go to Wyoming and twirl a rack of postcards, you may end up looking at a picture of a rabbit sprouting a pair of antlers. If you step into a diner, you may even get to see a jackalope in the flesh—or at least the head of one mounted on the wall.

On one level, it's pure bunk. The jackalopes on walls and postcards are nothing but taxidermic trickery—rabbits with pieces of antelope antler glued to their heads. But like many myths, the tale of the jackalope has a grain of truth buried at its core. Some real rabbits do indeed sprout horn-shaped growths.

In the early 1930s, a scientist at Rockefeller University named Richard Shope heard about horned rabbits while on a hunting trip. When he got back to New York, he arranged for a friend to catch one of these strange creatures and send him a piece of its horn. Shope wanted to take a careful look at it in his laboratory. He had reason to suspect that it was actually a tumor.

Shope got his suspicions from a colleague at Rockefeller named Francis Rous. Over 20 years earlier, in 1909, Rous was paid a visit by a Long Island chicken farmer. The farmer brought with her a Plymouth Rock hen with a worrisome growth protruding from its breast. She worried that it was some sort of infection that could spread to the rest of her flock.

Rous ground up the growth, mixed it in water, and passed it through a fine filter. He found, like Beijerinck before him, that he had created a contagious living fluid. He could use it to infect other chickens and cause them to produce growths as well. But when Rous looked at these growths under a microscope, they astonished him. They were tumors. Rous had discovered a cancer-causing virus. When Rous published his findings, most scientists greeted them with skepticism. The idea went against everything they thought they knew about both viruses and cancer. Their skepticism only grew

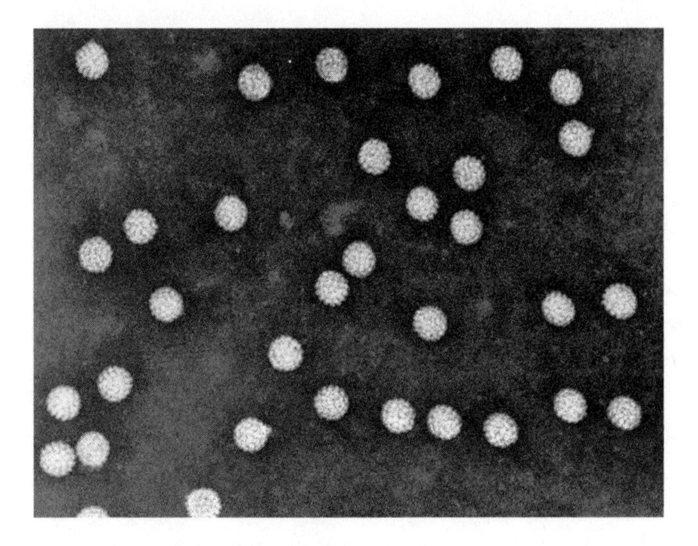

Human papillomaviruses (HPV) in suspension

as Rous tried to find cancer-causing viruses in other animals and failed to find any.

When Shope heard about jackalopes, he wondered if they were the animals Rous had been looking for. Once the jackalope horns arrived in New York, Shope followed Rous's experiment. He ground up the horns, mixed them in a solution, and then filtered the liquid through porcelain. The fine pores of the porcelain would let only viruses through. Shope then rubbed the filtered solution onto the heads of healthy rabbits. They grew horns as well. With this experiment, Shope did more than show that the horns contained viruses. He also proved that the viruses *created* the horns, crafting them out of infected cells.

Shope then passed on his rabbit tissue collection to Rous, who continued to work on it for decades. The rabbits presumably passed the virus to each other through physical contact, which explained why they produced tumors on their skin. Rous wondered what effect the viruses might have on other parts of the body. To find out, he injected virus-loaded liquid deep inside rabbits. Instead of harmless horns, the virus gave rise to aggressive cancers that killed the animals. For his research linking viruses and cancer, Rous won the Nobel Prize in Medicine in 1966.

The discoveries of Shope and Rous led scientists to look at growths on other animals. Cows sometimes develop monstrous lumps of deformed skin as big as grapefruits. Warts grow on mammals, from dolphins to tigers to humans. And on rare occasions, warts can turn people into human jackalopes.

In the early 1980s, an Indonesian boy named Dede Koswara began to develop warts on his knee. They soon spread to other parts of his body. Before long, they had overgrown his hands and feet, taking on the shape of giant claws. Eventually he could no longer work at a regular job and ended up as an exhibit in a freak show, earning the nickname "Tree Man." Reports of Dede began to appear in the news, and in 2007 doctors removed 13 pounds of warts from Dede's body. New growths returned, though, and so Dede had to undergo more surgery from time to time before his death in 2016 at age 45.

Dede's growths, along with all the others on humans and mammals, turned out to be caused by a single virus—the same kind that puts horns on rabbits. It's known as the papillomavirus, named for the papilla (*buds* in Latin) of virus-bearing cells that form during an infection.

At first, human papillomaviruses (HPV for short) didn't seem like a significant threat to public health. Cases like Dede's were very rare, and warts—while common—were generally harmless. But in the 1970s the German researcher Harald zur Hausen speculated that papillomaviruses might actually be a much bigger threat. He came to suspect they were the cause of cervical cancer, which kills over 300,000 women every year.

In women who develop cervical cancer, tumors grow on the cervix, the tissue joining the uterus to the vagina. As they become larger, they can damage the surrounding tissue, even tearing open a woman's intestines and causing fatal bleeding. When researchers studied the women who get cervical cancer, they noticed some odd patterns. It seemed to spread like a sexually transmitted disease. Nuns, for example, get cervical cancer much less often than other women. Some scientists had speculated that a virus spread during sex caused cervical cancer. Zur Hausen wondered if cancer-causing papillomaviruses were the culprit.

If this were true, zur Hausen reasoned, he ought to find virus DNA in cervical tumors. He gathered biopsies and sorted through their DNA. Stuck with the primitive scientific tools of the 1970s, he searched for years. Finally, in 1983 he discovered papillomavirus DNA lurking in some of the samples. For his efforts, zur Hausen shared the Nobel Prize for Physiology or Medicine in 2008.

Zur Hausen's discovery led generations of scientists to

study papillomaviruses and the remarkable way they hijack our cells. HPV specializes in infecting sheets of tissue in our body known as epithelia. Epithelial cells line our skin, our throats, and membranes across our bodies. The cells are stacked in layers, with the oldest layers on the top and the youngest at the bottom. As the top epithelial layer dies off and sheds dead cells, newer cells rising from below take its place.

To establish itself in our bodies, HPV infiltrates epithelia through nicks and works its way down to the deepest, youngest layers. Once it infects a new cell, it does not immediately kill it off, as rhinoviruses or influenza do. Instead, HPV uses a profoundly different strategy to thrive: the virus keeps its new host alive and even helps the cell multiply faster.

Speeding up a cell's division is no small feat, especially for a virus with just eight genes. It has to take over a series of biochemical reactions that are marvelously complex. A cell "decides" to divide in response to signals both from the outside and the inside, mobilizing an army of molecules to reorganize its contents. Its internal skeleton of filaments reassembles itself, pulling apart the cell's contents to two ends. At the same time, the cell makes a new copy of its DNA—over three billion "letters" all told, organized into 46 clumps called chromosomes apart. The cell must drag the two sets of chromosomes apart and build a wall between them. During this buzz of activity, supervising proteins monitor the progress. If these proteins sense that a cell is growing too quickly—perhaps thanks to a defective gene—they can trigger it to commit suicide. In doing so, they save us from cancer many times a day. The rate at which epithelial cells divide also changes as they rise to the surface. They slow down, channeling their resources into making a tough protein called keratin. Eventually they die, forming a protective shield on top of the more delicate epithelial cells underneath.

HPV utterly transforms a host cell. It disables the brakes that normally keep it growing at its proper pace. Knocking out the cell's cancer-monitoring proteins, the virus prevents them from sensing that something has gone wrong. Instead of dying as it rises, the infected cell keeps multiplying, creating a clump of virus-laden tissue. And when these cells approach the surface, they all suddenly manufacture huge numbers of new papillomaviruses. When they reach the surface, they rip open, spilling out the HPV.

This strategy has worked extraordinarily well for HPV. Most babies get colonized by the virus within a few days of birth. As people slough off dead skin, the virus drifts on dust particles and reaches new hosts. It also spreads easily through sex, with over 80 percent of sexually active adults acquiring at least one infection of HPV by this route. Infected cells typically reach the surface and die off before they can become troublesome.

Our own immune systems also help keep the virus in check. When cells get infected, they push fragments of viral proteins to their surface as an alarm signal. Immune cells passing by recognize this alarm and issue commands for the infected cells to commit suicide—destroying the viruses within them. As a result, the vast majority of people don't suffer at all from an HPV infection. Dede's plight demonstrates what happens when the immune system can't keep HPV in check. A rare genetic condition called epidermodysplasia verruciformis disables the communication network between epithelial cells and their immune cell minders. The infected cells multiply far faster than they die off, creating tree-like growths.

A far more common imbalance arises when HPV manages to establish itself in an epithelium for a long time. Instead of getting sloughed away after a few months, it creates an aggressive mass of infected cells that turns into a tumor. While the cervix is the most common tissue where HPV causes cancer, it's not the only one: it can also create tumors in the vagina, the penis, and the back of the throat.

In most of their hosts, however, papillomaviruses strike a peaceful balance—one that they've been striking for over 400 million years. Scientists have found hundreds of species of papillomaviruses infecting other animals—not just mammals, but birds, reptiles, and even fish. The genetic differences that have accumulated in the branches of the papillomavirus family tree are immense. These lines of evidence all suggest that our aquatic ancestors were already infected with the virus. As they diverged into different kinds of animals, papillomaviruses adapted to the evolving biology of their hosts.

In our own primate branch of the tree of life we can see the parallel tracks of virus and host evolution. About 40 million years ago, the ancestors of monkeys that live in Central and South America split off from the ancestors of monkeys and apes in Africa, Europe, and Asia. The papillomaviruses that infect living primates show the same split. Our papillomaviruses, for example, are more closely related to those of baboons in Kenya than those of howler monkeys in the Amazon.

About seven million years ago, our own lineage split off

from those of chimpanzees and other apes. Our ancestors became bipedal tool users wandering across much of Africa. About half a million years ago our lineage split in two. One branch moved out of Africa and evolved into Neanderthals and a similar group of humans called Denisovans. Back in Africa, our own species arose about 300,000 years ago. Only later, roughly 60,000 years ago, did the ancestors of non-African populations expand out of Africa, moving into Asia, Australia, and Europe. Neanderthals and Denisovans endured until about 40,000 years ago, coexisting with modern humans for thousands of years. Our DNA contains a record of that overlap. A small fraction of our genetic material matches the DNA of Neanderthals and Denisovans found in their fossils.

The best explanation for that match is that modern humans interbred with Neanderthals and Denisovans before their extinction. That explanation can also make sense of some of the puzzling patterns in the genes of our papillomaviruses. Certain strains of the virus are common today in people outside of Africa but rare among Africans. But these non-African viruses have peculiar mutations suggesting that they come from ancient lineages that date back long before humans expanded out of Africa. It looks as if modern humans acquired human papillomaviruses from sex with Neanderthals and Denisovans, and they've passed down these pathogens for tens of thousands of years.

Yet the most important feature of HPV's evolution remains a mystery: how it gained its power to cause deadly cancer in humans. The horns that rabbits develop from papillomaviruses may be striking, but they're benign. Outside of our own species, the virus rarely creates an aggressive tumor. What's more, only a few of the known strains of HPV account for most cancer in humans. Why they push epithelial cells toward cancer is a question yet to be answered.

Even if there's still a lot about HPV we don't understand, we already know enough to potentially do something once unthinkable: eradicate one kind of cancer with a vaccine. When it comes to preventing cancer, we're more accustomed to getting advice about not smoking or avoiding mutation-triggering chemicals. But once scientists uncovered the cancer-causing potential of HPV, they realized they could stop it. In the 1990s, researchers began developing a vaccine that contained proteins from the outer shell of HPV. Once vaccinated, people could mount a powerful immune attack against HPV before it could start pushing epithelial cells toward cancer. Clinical trials demonstrated that they provided complete protection against the two most common cancer-causing strains of the virus. In 2006, HPV vaccines were approved for use.

Australia quickly set up the first national program in 2007, and soon managed to vaccinate 70 percent of boys and girls of 12 or 13 years of age. Within three years, the number of precancerous cervical growths in Australian girls under 18 dropped by 38 percent. In Scotland, a 2019 study found that vaccination had reduced serious growths by 89 percent. HPV vaccines have proven so effective in countries with strong national programs, in fact, that they may be able to eradicate it within their borders in the coming years. Unfortunately many other countries—even including the wealthiest country on Earth, the United States—lag far behind. Thousands of women are developing virus-triggered cancer that could have been prevented.

In the future, other forms of cancer may get eliminated with vaccines of their own. Researchers have discovered that HPV is not unique in its ability to cause cancer. Liver-infecting hepatitis viruses can lead to liver cancer, for example, while Epstein-Barr virus can produce tumors in the esophagus and stomach. In total, scientists estimate, viruses cause 11 percent of all cancer cases—all potentially preventable by vaccines.

Even if every teenager gets vaccinated, however, cervical cancer may not disappear altogether. After all, the HPV vaccine only targets the two strains that cause most tumors. Scientists have identified 13 other cancer-causing strains of HPV, and there are likely others yet to be discovered. If vaccines decimate the two most successful strains, natural selection might well favor the evolution of other strains to take their place. Never underestimate the evolutionary creativity of a virus that can transform rabbits into jackalopes, or men into trees.

EVERYWHERE, IN ALL THINGS

The Enemy of Our Enemy

BACTERIOPHAGES AS VIRAL MEDICINE

By the beginning of the twentieth century, scientists had learned a few important things about viruses. They knew that viruses were infectious agents of unimaginably small size. They knew that certain viruses caused certain diseases, such as tobacco mosaic disease and rabies. But the young science of virology was still parochial. It focused mainly on the viruses that worried people most: the ones that made people sick, and the ones that threatened the crops and livestock they depended on. Virologists rarely looked beyond our little circle of experience. But during World War I, two physicians got a glimpse of the greater universe of viruses in which we live.

In 1915, Frederick Twort discovered this universe quite by accident. At the time, he was looking for an easier way to make smallpox vaccines. In the early 1900s, the standard vaccine for the disease contained a mild relative of smallpox called vaccinia. When doctors injected vaccinia into people, their immune systems made antibodies that could wipe out not just vaccinia, but smallpox, too. Twort wondered if he could grow large stocks of vaccinia by infecting cells he grew in Petri dishes.

His experiments ended in failure because bacteria contaminated his dishes and wiped out his cells. But Twort's frustration wasn't enough to blind him to something strange. He noticed that the carpets of bacteria that spread across his dishes became dotted with glassy spots. Under a microscope, Twort could see that they were full of dead microbes. He collected tiny drops from the glassy spots and transferred them to living bacterial colonies. In a matter of hours, new glassy spots formed, full of more dead bacteria. But when Twort added the drops to a different species of bacteria, no spots formed.

There were three explanations Twort could think of for what he was seeing. It might be some bizarre feature of the life cycle of the bacteria. Or perhaps the bacteria were committing suicide by producing deadly enzymes. The third possibility was the hardest to believe: maybe Twort had discovered a virus that kills bacteria.

Twort published his findings, listed the three possibilities, and left matters there. But two years later, a Canadian-born doctor named Felix d'Herelle independently made the same discovery and realized what he had actually discovered.

In 1917, d'Herelle was working as a military doctor, caring for French soldiers dying of dysentery. Dysentery, which leads to life-threatening diarrhea, is caused by bacteria known as *Shigella*. Doctors today can cure dysentery and other bacterial diseases with antibiotics, but these drugs would not be discovered until decades after World War I. D'Herelle was frustrated that he could do so little to help his patients. To better understand his enemy, d'Herelle examined their diarrhea.

He passed the stool of the soldiers through fine filters in order to trap *Shigella* and any other bacteria they contained. Only viruses and molecules could slip through. Once d'Herelle had produced this clear, bacteria-free fluid, he then mixed it with a fresh sample of *Shigella* bacteria and spread the mixture of bacteria and clear fluid in Petri dishes. The *Shigella* began to grow, but within a few hours d'Herelle noticed glassy spots starting to form in their colonies.

D'Herelle drew samples from those spots and mixed them with *Shigella* again. More glassy spots formed in the dishes. D'Herelle concluded that he was looking at miniature battlegrounds, where viruses were killing *Shigella* and leaving behind their translucent corpses.

At the time, this was a radical idea, since virologists only knew of viruses that infected animals and plants. D'Herelle decided his viruses deserved a name of their own. He dubbed them *bacteriophages*, meaning "eaters of bacteria." Today, they're known as *phages* for short.

Jules Bordet, a Nobel-prize-winning immunologist, read d'Herelle's report of his discovery and decided to look for more bacteriophages. Working in peacetime instead of war, Bordet did not use *Shigella* from sick soldiers. Instead, he chose a harmless favorite of lab workers, *Escherichia coli*. Following d'Herelle's example, Bordet poured *E. coli*–laden liquid through fine filters to isolate any phages it might contain. Then he mixed the filtered liquid with a second batch of *E. coli*. The second batch died, just as bacteria had died in d'Herelle's experiments.

But then Bordet took a step that d'Herelle had not. He decided to see what would happen if he added his filtered liquid to colonies of the first batch of *E. coli*—that is, the one he had filtered in the first place. If the liquid contained phages, they ought to kill those bacteria as well. To Bordet's surprise, no glassy spots formed. The first batch of *E. coli* was immune to whatever had killed the second. This surprise led Bordet to decide d'Herelle had been wrong: phages did not exist. Instead, Bordet argued, bacteria released toxic proteins that could kill other microbes, but not themselves.

D'Herelle fought back against Bordet, Bordet counterattacked, and the debate raged for years. It wasn't until the 1940s that scientists finally found the visual proof that d'Herelle was right. Training electron microscopes on fluid from those glassy spots, they discovered a strangely shaped virus, with a box-like shell sitting atop rod-like proteins that looked like the

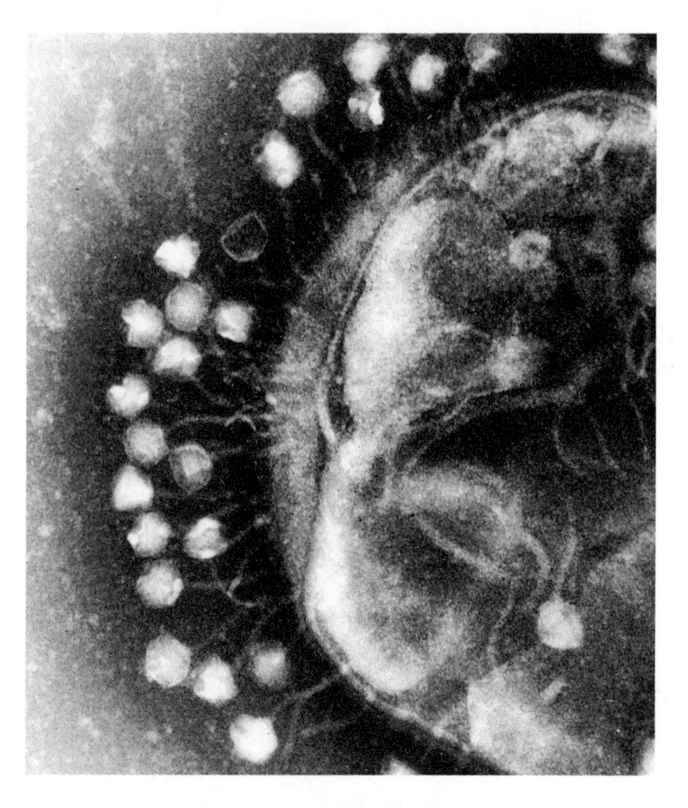

Bacteriophages attach to the surface of the host cell, a bacterium *Escherichia coli*

legs of spiders. Phages dropped onto the surface of *E. coli* like a lunar lander on the moon and then drilled into the microbe, squirting in their DNA.

Bordet wound up drawing the wrong conclusion from his experiment because he did not know that phages can have two profoundly different life cycles. D'Herelle's phages had to kill their hosts to replicate. They infected bacteria and immediately forced them to make new phages, which burst out of the microbes, leaving their ruptured remains behind. Virologists call such killers lytic phages.

Bordet, on the other hand, studied temperate phages—a kind of virus that can merge seamlessly into its host and leave it alive. Temperate phages treat bacteria much like human papillomaviruses treat our skin cells. When a temperate phage infects a host, the microbe inserts the virus's genes into its own DNA. The infected bacterium continues to grow and divide, making new virus genes along with its own. It is as if the virus and its host become one.

But temperate phages remain a lurking threat. If infected bacteria suddenly experience some kind of stress, that signal triggers them to read the genes of their embedded phage and make new viruses. The phages burst out of the cell and seek out vulnerable new hosts. But they can only invade microbes that are not already carrying a temperate phage. Bordet's experiment failed because his original stock of bacteria had a viral immunity.

D'Herelle did not wait for the debate over phages to end before he began trying them out as a cure for dysentery. If he gave patients an extra supply of phages, they might be able to wipe out all the bacteria and clear the infection. Before he could test this hypothesis, d'Herelle first needed to be sure phages were safe. He filtered fluid laced with *Shigella* to create a batch of phages, and then he drank them "without detecting the slightest malaise," as he later wrote. D'Herelle then injected the phage-loaded fluid into his skin, again with no ill effects.

Now confident that phages were safe, d'Herelle began to use his "phage therapy" on sick patients. He reported that they helped people recover from dysentery. He tried it out on other bacterial diseases, like cholera, and reported more success. When four passengers on a French ship in the Suez Canal came down with bubonic plague, d'Herelle gave them phages. All four victims recovered.

D'Herelle's discovery of bacteriophages had made him famous within scientific circles, but now his cures made him a celebrity. The American writer Sinclair Lewis made him the basis for his 1925 best-selling novel, *Arrowsmith*, which Hollywood turned into a movie in 1931. Meanwhile, d'Herelle developed phage-based drugs sold by the company that's now known as L'Oréal. Droves of customers used his phages to treat skin wounds and to cure intestinal infections.

But the phage craze did not last long. In the 1930s, researchers discovered the first antibiotics—molecules produced by fungi and bacteria that could wipe out infections. Doctors eagerly switched to these inert, reliable chemicals. Antibiotics soon proved to be staggeringly effective, and reliably so. The market for phage therapy dried up, and most scientists saw little reason to investigate it further.

Yet d'Herelle's dream did not vanish. On a trip to the Soviet Union in the 1920s, while he was still a medical icon, he met scientists who wanted to set up an entire institute to conduct research on phage therapy. In 1923 d'Herelle helped some of them establish the Eliava Institute of Bacteriophage, Microbiology, and Virology in Tbilisi, which is now the capital of the Republic of Georgia. At its peak, the institute employed 1,200 people and produced tons of phages a year. During World War

II, the Soviet Union shipped phage powders and pills to the front lines, where they were dispensed to infected soldiers. The Eliava team even ran a massive clinical trial in 1963 to demonstrate that phage therapy actually worked. Around Tbilisi, they gave children on one side of each street a pill containing phages. The children on the other side got a sugar pill. All told, they enrolled 30,769 children in the study and then followed them for 109 days. Among the children who took the sugar pill, 6.7 out of every 1,000 got dysentery. Among the children who took the phage pill, that figure dropped to 1.8 per 1,000. Taking phages caused a 3.8-fold decrease in a child's chance of getting sick.

If such a study had taken place in the West, it might have led some scientists to give phage therapy another look. But thanks to the wall of secrecy that the Soviet government erected around the country's science, few people became aware of it outside of Georgia. Only after the Soviet Union fell in 1989 did Westerners learn the full scope of the remarkable work going on in Tbilisi. By then, infectious disease experts were finally willing to seriously consider alternatives to antibiotics. The wonder drugs were starting to fail, as antibiotic resistance became more common. Doctors found that their safest, most reliable antibiotics could no longer stop infections. They had to resort to backups that were more expensive and sometimes had dangerous side effects.

By the 1990s, a number of researchers were giving phage therapy a serious look. But they could recognize some big challenges that blocked its path to the clinic. Phages come in a vast diversity of species and strains, for example, each finely adapted to a particular bacterial host. Even if a phage proved effective against one strain of a pathogen, it might fail against others.

Skeptics also worried that phages, like antibiotics, would succumb to resistance as well. In the 1940s, the microbiologists Salvador Luria and Max Delbruck observed bacteria evolving resistance to phages before their own eyes. When they laced a dish of *E. coli* with phages, most of the bacteria died, but a few clung to existence and then later multiplied into new colonies. Further research revealed that those survivors had acquired mutations that allowed them to resist the phages. The resistant bacteria then passed on their mutated genes to their descendants. Critics have argued that phage therapy would turn our bodies into Petri dishes, where bacteria could evolve resistance to phages.

In the twenty-first century, phage therapy researchers have overcome some of these worries. It is certainly true that phages are picky about their hosts, but that doesn't rule out using phage therapy to cure a wide range of infections. Scientists at the Eliava Institute have developed a dressing for wounds that is impregnated with half a dozen different phages, for example, capable of killing the six most common kinds of bacteria that infect skin wounds. Researchers are also creating collections of phages, against which each patient's bacteria can be tested to find one that works against it.

As scientists find new phages, they're discovering species that can deliver new attacks against bacteria. Ben Chan, a research scientist at Yale University, and his colleagues have discovered a phage that slips into bacteria through a pump in their surface. That pump just so happens to be the one that the bacteria use to push antibiotics out of their interior before they can cause any harm. Bacteria can evolve stronger resistance to antibiotics by making more of these pumps.

Chan and his colleagues tested out their new phage in a dish of bacteria. If they exposed the bacteria to the phage, the microbes evolved fewer pumps to make it harder for the phage to infect them. But with fewer pumps, they became more vulnerable to antibiotics. Their study suggests that phages and antibiotics together could trap bacteria in an evolutionary conflict. Soon afterward, Chan and his colleagues gave this combination to a man with a chronic heart infection of resistant bacteria. The bacteria became vulnerable to antibiotics, and he recovered.

Of course, a trial on a single patient is no more proof that phage therapy is safe and effective than it was in d'Herelle's day. But Chan and other researchers are treating more people to see if phage therapy can help, and other researchers have launched clinical trials. Governments are now trying to make this research easier by developing regulations more fitting for viruses than for drugs. Over a century after d'Herelle first encountered bacteriophages, these viruses may finally be ready to become a part of modern medicine.

The Infected Ocean

HOW MARINE PHAGES RULE THE SEA

Some great discoveries seem at first like terrible mistakes.

In 1986 a graduate student at the State University of New York at Stony Brook named Lita Proctor decided to see how many viruses there are in seawater. At the time, the general consensus was that there were hardly any. The few researchers who had bothered to look for viruses in the ocean had found only a scarce supply. Most experts believed that the majority

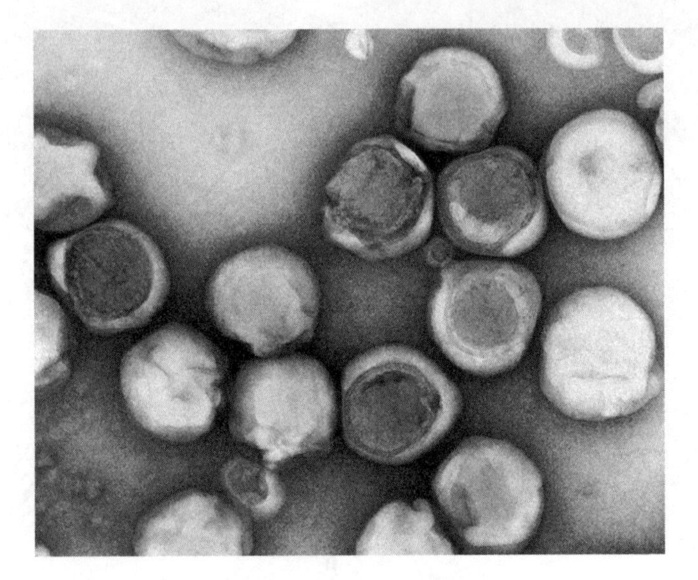

Emiliania huxleyi viruses infect ocean algae
(viruses shown here in suspension)

of the viruses they did find in seawater had actually come from sewage and other sources on land.

But over the years, some evidence emerged that didn't fit neatly into the consensus. A marine biologist named John Sieburth had published a photograph of a marine bacterium erupting with new viruses, for example. Proctor decided to launch a systematic search to see just how many viruses were in the ocean. She traveled to the Caribbean and to the Sargasso Sea, scooping up seawater along the way. Back on Long Island, she carefully extracted the biological material from the seawater, which she coated with metal so that it would show up under the beam of an electron microscope. When Procter finally looked at her samples, she beheld a world of viruses. Some floated freely, while others were lurking inside bacteria. Based on the number of viruses she found in her samples, Proctor estimated that every liter of seawater contained up to 100 billion viruses.

Proctor's figure far exceeded previous estimates. But when other scientists followed up on her work and carried out their own surveys, they ended up with similar figures. They found viruses lurking in deep-sea trenches and locked in Arctic sea ice. They came to agree there are somewhere in the neighborhood of 10,000,000,000,000,000,000,000,000,000,000 viruses in the ocean.

It is hard to find a point of comparison to make sense of such a huge number. There are 100 billion times more viruses in the oceans than the grains of sand on all the world's beaches. If you put the viruses of the oceans together on a scale, they would equal the weight of 75 million blue whales (there are less than 10,000 blue whales on the entire planet). And if you lined up all the viruses in the ocean end to end, they would stretch out 42 million light-years.

These numbers don't mean that a swim in the ocean is a death sentence. Only a minute fraction of the viruses in the ocean can infect humans. Some marine viruses infect fishes and other marine animals. But their most common targets are bacteria and other single-celled microbes. Microbes may be invisible to the naked eye, but collectively they dwarf all the ocean's whales, its coral reefs, and all other forms of marine life. And just as the bacteria that live in our bodies are attacked by phages, marine microbes are attacked by marine phages.

When Felix d'Herelle discovered the first bacteriophage in French soldiers in 1917, many scientists refused to believe that such a thing actually existed. A century later, it's clear that d'Herelle had found the most abundant life form on Earth. What's more, marine phages have a massive influence on the planet. They influence the ecology of the world's oceans. They leave their mark on Earth's global climate. And they have been playing a crucial part in the evolution of life for billions of years. They are, in other words, biology's living matrix.

Marine phages are powerful because they are so infectious. They invade a new microbe host 100 billion trillion times a second, and they kill between 15 and 40 percent of all bacteria in the world's oceans every single day. Out of these dying hosts come new marine phages. Every liter of seawater generates up to 100 billion new viruses every day.

Their lethal efficiency keeps their hosts in check, and we humans often benefit from their deadliness. Cholera, for example, is caused by blooms of waterborne bacteria called *Vibrio*. But *Vibrio* are host to a number of phages. When the population of *Vibrio* explodes and triggers an epidemic, the phages multiply. The virus population rises so quickly that it kills *Vibrio* faster than the microbes can reproduce. The bacterial boom subsides, and the cholera epidemic fades away.

It's not just new viruses that spill out of a dead microbe. Its organic carbon and other molecules are liberated as well. Each year, ocean viruses unlock billions of tons of carbon, and this massive injection has a planet-wide impact. It acts like a fertilizer, stimulating the growth of vast numbers of new microbes, some of which support the ocean's food web. That web might very well be smaller if viruses were not spurring this growth. Some of the liberated carbon doesn't get taken up by microbes; instead, it sinks down to the bottom of the ocean. The molecules inside a microbe are sticky, and so once a virus rips open a host, the glue-like molecules that spill out snag other carbon molecules, creating a vast blizzard of underwater snow tumbling to the sea floor.

The hosts of the ocean's viruses have responded to this threat by evolving all manner of defenses. But the viruses have evolved ways to override them. Since each species follows an evolutionary escape route of its own, this race has helped produce a staggering diversity of marine viruses. Lita Proctor could not fathom how many different kinds of viruses she was discovering when she started her search. Looking through a microscope, she could count them, but she only saw a limited number of shapes—balls, cylinders, and the like. But in the world of viruses, appearances are deceiving. Rhinoviruses and polioviruses may look like nearly identical spheres, but the former cause mild colds while the latter can paralyze or kill.

Starting in the early 2000s, virologists discovered how to go beyond appearances, by looking directly at virus genes. They would gather a sample—be it sea water or dirt or the insides of a bumblebee—and filter out all but the viruses from it. Then they would extract the genetic material from the viruses and read their sequences. In some cases, those sequences matched some well-known species or strain of virus. Very often, however, it didn't. Wherever scientists looked, they found staggering diversity of viruses. Even in our own bodies surprises were waiting. In 2014, a team of scientists led by Bas Dutilh discovered a new phage in human feces they dubbed crAssphage (short for "cross assembly," a method for piecing together virus gene sequences). Researchers soon found many more types of crAssphage-like viruses, which made up to 90 percent of all the viruses in people's bodies. Yet they had gone unnoticed for a century after Felix d'Herelle's discovery of phages.

It is in the oceans where the true scope of the virosphere has become clear. Matthew Chapman, a virologist at Ohio State University, and his colleagues analyzed genetic material in sea water collected on a scientific voyage around the world. In 2016, they reported over 15,000 new species of viruses. By comparison, there are only 6,400 species of mammals. Chapman and his colleagues thought they had pretty much figured out the diversity of viruses in the sea, but just to be sure, they continued to collect more water and invented new ways to search it for viral genes. In 2019, they reported finding a total of 200,000 species. Yet they left the vast majority of the ocean untouched. All told, some researchers have estimated the Earth may be home to 100 trillion species of viruses—most of which can be found at sea.

Viruses evolve this diversity thanks to the peculiar way they multiply. When cells get infected, they create many new viruses, but they do so sloppily. The genes of the new viruses are rife with copying mistakes. While most of these mutations disable viruses, some of them give viruses an evolutionary edge, letting them infect hosts more effectively. If two kinds of viruses infect a cell at once, they may shuffle their genes together. Viruses can even end up with some of their host's own genes, which they can then deliver into new hosts. By one estimate, ocean viruses transfer a trillion trillion genes between the genomes of their hosts every year.

Thanks to gene borrowing, viruses may be responsible for a lot of the world's oxygen. Much of the oxygen in the atmosphere is produced by photosynthetic microbes in the oceans. Some of the viruses that infect them carry their own genes for photosynthesis. When they invade, the viruses take charge of harvesting light. By one rough calculation, 10 percent of all the photosynthesis on Earth is carried out with virus genes. Breathe 10 times, and one of those breaths comes to you courtesy of a virus.

This shuttling of genes has had a huge impact not just on Earth today, but throughout the history of life. It was in the oceans that life got its start, after all. The oldest traces of life are fossils of marine microbes dating back almost 3.5 billion years. It was in the oceans that multicellular organisms evolved; their oldest fossils date back to about 2 billion years ago. In fact, our own ancestors did not crawl onto land until about 400 million years ago. Viruses don't leave behind fossils in rocks, but they do leave marks on the genomes of their hosts. Those marks suggest that viruses have been around for billions of years.

Scientists can determine the history of genes by comparing the genomes of species that split from a common ancestor that lived long ago. Those comparisons can, for example, reveal genes that were delivered to their current host by a virus that lived in the distant past. Scientists have found that all living things have mosaics of genomes, with hundreds or thousands of genes imported by viruses. As far down as scientists can reach on the tree of life, viruses have been shuttling genes. Darwin may have envisioned the history of life as a tree. But the history of genes, at least among the ocean's microbes and their viruses, is more like a bustling trade network, its webs reaching back billions of years.

Our Inner Parasites

ENDOGENOUS RETROVIRUSES AND OUR VIRUS-RIDDLED GENOMES

The idea that a host's genes could have come from viruses is almost philosophical in its weirdness. We like to think of our genomes as our ultimate identity. The fact that bacteria have acquired much of their DNA from viruses raises baffling questions. Do they have a distinct identity of their own? Or are they just hybrid Frankensteins, their clear lines of identity blurred away?

At first, it was possible to cordon off this puzzle from our own existence, treating it purely as a question about microbes. The presence of viral genes was merely a fluke of "lower" life forms. But today we can no longer find such comfort. If we look inside our own genome, we now see viruses. Thousands of them.

It took many decades for scientists to recognize our inner viruses, and at the start of this journey was Francis Rous's Plymouth Rock chicken. That sick hen spurred Rous's fifty-year investigation of cancer-causing viruses. Rous and other researchers came to find many different viruses that can produce tumors. Studying rabbits, for example, Rous made pioneering studies of papillomaviruses. His chicken, it turned out, was infected with yet another species that came to bear his name: Rous sarcoma virus.

Later generations of scientists studied Rous sarcoma virus, hoping to unlock some of cancer's secrets. In the process, they discovered that the virus has a remarkable way of replicating. Rous sarcoma virus encodes its genes in single-stranded RNA. When it infects a chicken's cell, it makes a copy of its genes in double-stranded DNA. And then that viral DNA gets inserted into the host's genome. When the host cell divides, the researchers found, it copies the virus's DNA along with its own. Under certain conditions, the cell is forced to produce new viruses—complete with genes and a protein shell—which can then escape to infect a new cell. Chickens develop tumors if Rous sarcoma virus genes are accidentally inserted in the wrong location in their genome. The virus's genes cause nearby host genes to switch on that should have stayed off, triggering the cell into a runaway growth. In the

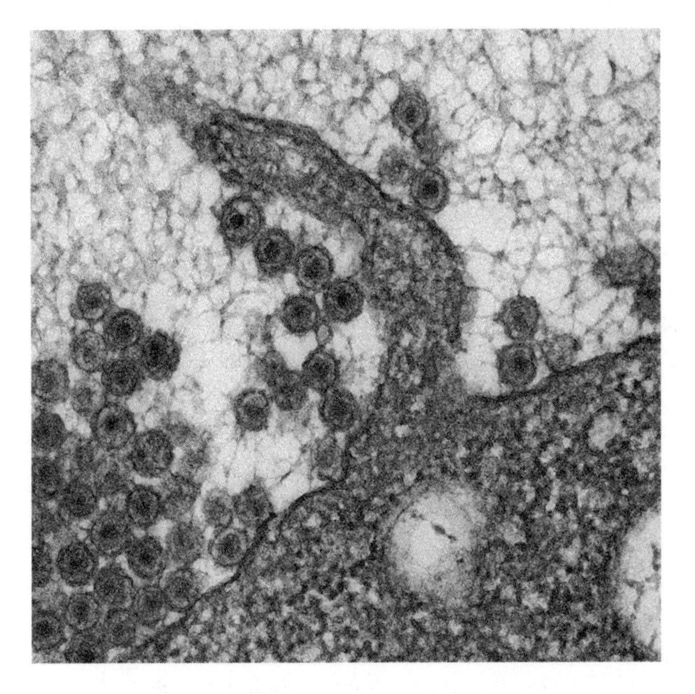

Avian leukocyte viruses bud from a human white blood cell

1960s, researchers came to recognize that Rous sarcoma virus is far from unique. Many other viruses, collectively known as retroviruses, insert their genes into host genomes in the same way.

Robin Weiss, a virologist then working at the University of Washington, developed a curiosity about one retrovirus in particular. It was known as avian leukosis virus, a close relative of Rous sarcoma virus. Weiss was puzzled by the results of tests on chickens for the presence of the virus. The tests involved screening the blood of birds for the virus's proteins. Sometimes the viral proteins turned up in chickens that were healthy and never developed cancer. Stranger still, when those healthy chickens produced chicks, the chicks were born with viral proteins, too.

Thinking back to how retroviruses insert their genes into their host genomes, Weiss wondered if they could be passed down through generations of chickens. To try to flush the virus out of its hiding place, Weiss and his colleagues grew cultures of cells from healthy chickens that produced the viral protein. They bathed the cells in mutation-triggering chemicals and battered them with radiation—the kinds of assaults that trigger retrovirus genes to awaken and make new viruses.

Just as they had suspected, the mutant cells started to churn out new avian leukosis viruses. In other words, these healthy chickens were not simply infected with avian leukosis virus in *some* of their cells; the genetic instructions for making the virus were implanted in *all* of their cells, and the birds passed those instructions down to their descendants.

Weiss and his colleagues soon found that these hidden viruses were not limited to just one oddball breed of chicken. They found avian leukosis virus embedded in many breeds, raising the possibility that the virus was an ancient component of chicken DNA. To see just how long ago avian leukosis viruses infected the ancestors of today's chickens, Weiss and his colleagues traveled to the jungles of Malaysia. There they trapped red jungle fowl, the closest wild relatives of chickens. The red jungle fowl carried the same avian leukosis virus, Weiss found. On later expeditions, he found that other species of jungle fowl lacked the virus.

Out of this research on avian leukosis virus emerged a hypothesis for how it had merged with chickens. Thousands of years ago, the virus plagued the common ancestor of domesticated chickens and red jungle fowl. It invaded cells, made new copies of itself, and infected new birds, leaving tumors in its wake. But in at least one bird, something else happened. Instead of giving the bird cancer, the virus was kept in check by the bird's immune system. As it spread harmlessly through the bird's body, it infected the chicken's sexual organs. Infected eggs or sperm could give rise to infected chicken embryos.

As an infected embryo grew and divided, all of its cells also inherited the virus DNA. When the chick emerged from its shell, it was part chicken and part virus. And with the avian leukosis virus now part of its genome, the chicken passed down the virus's DNA to its own offspring. The virus remained a silent passenger from generation to generation for thousands of years. But under certain conditions, the virus could reactivate, create tumors, and spread to other birds.

Scientists recognized that avian leukosis viruses was in a class of its own. They called it an endogenous retrovirus—endogenous meaning *generated within*. They soon found endogenous retroviruses in other animals. In fact, the viruses lurk in the genomes of just about every major group of vertebrates, from fish to reptiles to mammals. Some of the newly discovered endogenous retroviruses turned out to cause cancer like avian leukosis virus does. But many did not. They were crippled with mutations that robbed them of the ability to make new viruses that could escape their host cell. These hobbled viruses could still make new copies of their genes, though, which were reinserted back into their host's genome. And scientists also discovered other endogenous retroviruses that were so riddled with mutations that they could no longer do anything at all. They had become nothing more than baggage in their host's genome.

Scanning the human genome, researchers began finding endogenous retroviruses there as well. As far as scientists can tell, none of them are active. But Thierry Heidmann, a researcher at the Gustave Roussy Institute in Villejuif, France, and his colleagues have discovered that they can transform this genetic baggage back into full-blown viruses. Heidmann was studying an endogenous retrovirus when he noticed that different people had slightly different versions. These differences presumably arose after a retrovirus became trapped in the genomes of ancient humans. In their descendants, mutations struck different parts of the virus's DNA.

Heidmann and his colleagues compared the variants of the virus-like sequence. It was as if they found four copies of a play by Shakespeare transcribed by four careless clerks. Each clerk might make his own set of mistakes, and sometimes one of Shakespeare's words ended up as four different misspelled versions—*wheregore, sherefore, whorefore, wherefrom*. By comparing all four versions, a historian could figure out that the original word was *wherefore*.

Using this method, Heidmann and his fellow scientists were able to use the mutated retroviruses in living humans to determine the sequence of the original one. They then synthesized a piece of DNA with that sequence and inserted it into human cells they reared in a culture dish. Out of the cells sprang new viruses that could infect other cells. In other words, the original sequence of the DNA had been a living, functioning virus. In 2006, Heidmann named the virus Phoenix, for the mythical bird that rose from its own ashes.

The Phoenix virus probably infected our ancestors within the last million years. But we also carry far older viruses, too. We know this because scientists have found the same viruses lurking both in our genome and in other species. In one study, Adam Lee, a virologist at Imperial College London, and his colleagues discovered an endogenous retrovirus called ERV-L in the human genome. They then found it in many other species ranging from horses to aardvarks. When Lee and his colleagues drew out an evolutionary tree of the virus, it mirrored the tree of the virus's hosts. It appears that this endogenous retrovirus infected the common ancestor of all mammals with placentas, which lived over 100 million years ago. Today, that virus lingers on, in armadillos and elephants and manatees. And in us.

As an endogenous retrovirus gets trapped in its host, it can still make new copies of its DNA, which get inserted back into its host's genome. Over the millions of years that endogenous retroviruses have been invading our genomes, they've accumulated to a staggering extent. Each of us carries almost 100,000 fragments of endogenous retrovirus DNA in our genome, making up about 8 percent of our DNA. To put that figure in perspective, consider that the 20,000 protein-coding genes in the human genome make up only 1.2 percent of our DNA.

Scientists have also found millions of smaller pieces of DNA that also get copied and inserted back in the human genome. It's possible that many of these pieces are tiny relics of endogenous retroviruses. Over millions of years, evolution stripped them down to the bare essentials required for copying DNA. Our genomes, in other words, are awash in viruses.

Most of this viral DNA has lost its ability to do anything thanks to millions of years of mutations. But our ancestors commandeered some viral genes for their own benefit. In fact, without these viruses, none of us today would have been born.

In 1999, Jean-Luc Blond and his colleagues discovered a human endogenous retrovirus they dubbed HERV-W. And they were surprised to discover that one of the genes could still produce a protein. The protein, called syncytin, turned out to have a very precise, very important job to do—not for the virus, but for its human host. It could be found only in the placenta.

The cells in the outer layer of the placenta made syncytin in order to join together, so that molecules could flow between them. Scientists discovered that mice, like humans, made syncytin, a discovery that allowed them to run experiments to understand how the protein worked. When they deleted the gene for syncytin, mouse embryos never survived to birth. The viral protein was essential for drawing nutrients from their mother's bloodstream.

Scientists looked for syncytin in other placental mammals, and they found it. But they came to realize that different species carried different forms of syncytin. Thierry Heidmann, who has discovered many of these syncytin proteins, has proposed a scenario to make sense of all these viruses in the placenta. Over 100 million years ago, an ancestral mammal was infected by an endogenous retrovirus. It harnessed the first syncytin protein and evolved the very first placenta. Over the course of millions of years, that original placental mammal gave rise to many lineages of descendants. And they continued to be infected with endogenous retroviruses. In some cases, the new viruses had syncytin genes of their own, which produced superior proteins for the placenta. Different lineages of mammals—rodents, bats, cows, primates—swapped one viral protein for another.

In our most intimate moment, as new human life emerges from old, viruses are essential to our survival. There is no us and them—just a gradually blending and shifting mix of DNA.

THE VIRAL FUTURE

The Young Scourge

HUMAN IMMUNODEFICIENCY VIRUS AND THE ANIMAL ORIGINS OF DISEASES

Every week, the Centers for Disease Control and Prevention publishes a thin newsletter called *Morbidity and Mortality Weekly Report*. The issue that appeared on July 4, 1981, was a typical assortment of the ordinary and the mysterious. Among the mysteries that week was a report from Los Angeles, where doctors had noticed an odd coincidence. Between October 1980 and May 1981, five men were admitted to hospitals around the city with the same rare disease, known as pneumocystis pneumonia.

Pneumocystis pneumonia is caused by a common fungus called *Pneumocystis jiroveci*. Most people inhale its abundant spores at some point during their childhood, but their immune system normally kills them off and produces antibodies that ward off any future infection. In people with weak immune systems, on the other hand, *P. jiroveci* runs rampant. The lungs fill with fluid and become badly scarred. Pneumocystis pneumonia leaves its victims struggling to inhale enough oxygen to stay alive.

The five Los Angeles patients did not fit the typical profile of a pneumocystis pneumonia victim. They were young men who had been in perfect health before they came down with the disease. Commenting on the report, the editors of *Morbidity and Mortality Weekly Report* speculated that the puzzling symptoms of the five men "suggest the possibility of a cellular-immune dysfunction."

Little did they know that they were publishing the first observations of what would become one of the deadliest viral epidemics in history. The five Los Angeles men's immune systems had been wiped out by a virus that would later be dubbed human immunodeficiency virus (HIV for short). Researchers would discover that HIV had been secretly infecting people for well over 50 years. It went on to explode into a worldwide catastrophe. By 2019, it had infected an estimated 75.7 million people, killing 32.7 million of them.

HIV's death toll is all the more terrifying because, as viruses go, it's hard to catch. It doesn't waft on the air or stick to surfaces. Only certain bodily fluids such as blood and semen can transmit the virus. The most common routes of infection included unprotected sex, childbirth, and sharing contaminated needles.

Once HIV gets into a person's body, it boldly attacks the immune system itself. It grabs on to certain kinds of immune cells known as CD4 T cells and fuses their membranes, like a tiny soap bubble merging into a bigger one. HIV is a retrovirus, which means that it inserts its genetic material into the cell's DNA. From there, it manipulates the cell to make new viruses that can escape to infect other CD4 T cells.

At first the virus's numbers skyrocket into the billions. Once the immune system recognizes infected CD4 T cells, however, it starts to kill them, driving the virus's population down. To the infected person, this battle feels like a mild flu. The immune system exterminates most of the HIV, but a small fraction of the virus manages to survive. The CD4 T cells in which the surviving HIV hides continue to grow and divide. From time to time, an infected CD4 T cell creates a batch of new HIV, which blast out to infect new cells. The immune system attacks these new waves of viruses each time they surge through the body.

Eventually this cycle of attack and evasion exhausts the immune system. It may take a year for it to collapse, or over a decade. Once the immune system fails, people can no longer defend themselves against diseases that would never be able to harm a person with a healthy immune system—diseases like pneumocystis pneumonia. This weakened condition came to be known as acquired immunodeficiency syndrome, or AIDS.

In 1983, two years after the first AIDS patients came to light, French scientists isolated HIV from a patient with AIDS

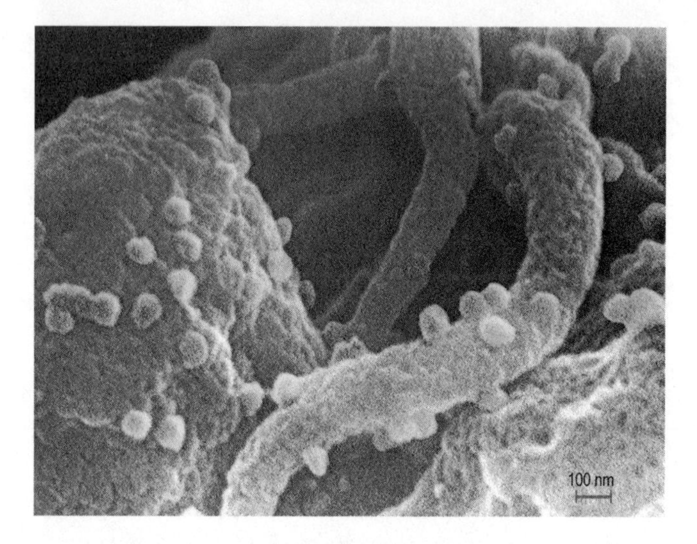

Human immunodeficiency viruses (HIV) on the
surface of a CD4 white blood cell

for the first time, and further research confirmed HIV as the cause of the disease. Meanwhile, AIDS was proving not to be an obscure condition among a few men in Los Angeles. New cases emerged across the United States and abroad. Other great scourges, such as malaria and tuberculosis, are ancient enemies, which have been killing people for thousands of years. Yet it only took a few years for HIV to go from utter obscurity to a global scourge. Here was an epidemiological mystery that would take scientists three decades to solve.

The first clues came from sick monkeys.

At primate research centers around the United States, pathologists noticed a number of animals with an AIDS-like condition. They wondered if the monkeys were getting infected with an HIV-like virus. In 1985, scientists at the New England Regional Primate Research Centers tested that idea by mixing antibodies for HIV into the serum of sick monkeys. If there was an HIV-like virus in their blood, the antibodies might stick to them. Their hunch was right: they were able to fish out retroviruses from the monkey serum. The new retroviruses came to be known as simian immunodeficiency virus, or SIV for short. Some of the SIVs scientists discovered were only distantly related to HIV, while others were closer kin.

Of course, scientists had to be careful about drawing too many conclusions from finding viruses in captive monkeys. SIVs might be rare or nonexistent outside of zoos and labs. But searching for SIVs infecting wild primates was no simple task, since the animals would not submit easily to a blood draw. Primatologists and virologists figured out how to isolate virus genes from urine and feces the primates left behind on leaves or the forest floor.

These expeditions have revealed that over half of all species of monkeys and apes in Africa carry their own strains of SIV. By comparing the genes of these strains, evolutionary biologists have drawn their family tree. They have determined that all SIVs descend from an ancestral retrovirus that infected an African monkey millions of years ago. At first it spread in this original monkey species through sex. But it then jumped to other species—perhaps when monkeys fought for territory and drew blood. When different strains of SIV ended up in the same cell, they could shuffle their genes to together to make entirely new strains.

As scientists have isolated strains of HIV in patients

around the world, they've added their branches to this viral family tree. The tree has revealed, as scientists suspected in the 1980s, that HIV evolved from SIV. But HIV did not have a single origin. It arose at least 13 separate times.

The first hint of these many beginnings came about in 1989, when scientists isolated an extremely HIV-like virus from a monkey called a sooty mangabey. They discovered that the virus, which was dubbed SIVsm, was more closely related to some HIV strains than others. At the time, scientists recognized two main kinds of HIV, which they called HIV-1 and HIV-2. HIV-1 is found around the world, while HIV-2 is limited for the most part to West Africa, where it causes a far less aggressive form of AIDS. The researchers who found SIVsm discovered that it was more closely related to HIV-2 than HIV-1. In later years, they found more strains of SIVsm. Some were more closely related to certain strains of HIV-2 than others. The best explanation for this evolutionary pattern is that SIVsm jumped from sooty mangabeys into humans at least nine times. No one witnessed those nine leaps, but we can be fairly certain how they happened. In West Africa, many people keep sooty mangabeys as pets. It's also common for hunters to kill the monkeys and sell their meat. The virus gained the opportunity to move from sooty mangabeys to people whenever their blood made contact—when a monkey bit a hunter, for example, or when a butcher prepared its meat. SIVsm could then infect human cells, replicate, and adapt to its new host species.

None of these jumps have been very successful. HIV-2 only replicates slowly and does a bad job of transmitting from person to person. Together, the nine strains of HIV-2 infect only an estimated one or two million West Africans. Researches who have closely examined the way HIV-2 infects human cells have found some possible explanations for why it has fared so poorly. When new HIV-2 viruses are ready to escape a cell, for example, the cell makes a lasso-like protein called tetherin that snags them and prevents them from leaving.

HIV-1 has proven far more successful as a human virus, but its origins have taken much longer to decipher. In 1999, researchers studying chimpanzees discovered a new SIV that was closer to all HIV-1 strains than to HIV-2. They named it SIVcpz. As they discovered more strains of SIVcpz, they figured out that it evolved from a blend of three different strains of SIV, each in a different monkey. The chimpanzees likely picked up the viruses as they hunted monkeys as prey. For millions of years, SIVcpz circulated from chimpanzee to chimpanzee, evolving into different strains across its range throughout central Africa.

Scientists have found that SIVcpz evolved into HIV-1 on four separate occasions. In two cases, the virus jumped directly from chimpanzees to humans. In two others, the virus spread to gorillas, which then passed it along. Three of these jumps produced only rare strains of HIV-1. But the fourth—which originated in chimpanzees that live in Cameroon—produced a lineage of viruses called HIV-1 Group M that today accounts for 90 percent of all HIV-1 infections. (M is short for *main.*)

It appears that HIV-1 Group M evolved into a more successful human parasite than the other versions of HIV. Scientists suspect that part of its success has to do with its interaction with tetherin. Unlike other strains of HIV, it evolved the ability to cut this molecular lasso, allowing the virus to slip more easily out of our cells.

While scientists didn't discover HIV until the 1980s, they suspected that these jumps had taken place long before. To pin down the timing of the virus's history, scientists searched for HIV in samples of blood and tissue taken from patients long before the discovery of HIV. In 1998, David Ho and his colleagues at Rockefeller University discovered HIV-1 Group M in a 1959 sample. It came from a patient in Kinshasa, the capital of the African country of Zaire (now the Democratic Republic of Congo). In 2008, Michael Worobey and his colleagues at the University of Arizona discovered another sample of HIV-1 Group M in a tissue sample from another pathology collection in Kinshasa, dating back to 1960.

To reach further back, scientists have extracted the history encoded in HIV's genes. As the virus replicates, it accumulates mutations at a clock-like rate, piling up like sand in an hourglass. By measuring the height of this genetic sand pile, scientists can estimate how much time had passed. Using this method, scientists found that the HIV-1 Group M samples from Kinshasa both originated in the early 1900s.

All this evidence points to an explanation for how HIV-1 got its start—or rather, its starts. For centuries, hunters in Cameroon killed both chimpanzees and gorillas for meat. And from time to time they would become infected with SIVcpz from the apes. But these hunters, living in relative isolation before the twentieth century, were a dead end for the viruses. Some people recovered from SIVcpz infections because their immune systems could vanquish the poorly adapted hosts. In other cases, the viruses were extinguished with the death of their host, unable to reach a new one.

Africa started undergoing dramatic changes in the early

1900s that gave SIV new opportunities to spread into humans. Commerce along the rivers allowed people to move from villages to towns, bringing their viruses with them. The colonial settlements in central Africa began to expand to cities of 10,000 people or more, giving the virus more opportunities to spread from host to host. HIV-1 Group M may have remained rare in humans for years. At some point, it gained adaptations that let it spread more successfully from person to person. It also got a lucky break. Somehow the virus traveled in the mid-1900s to Kinshasa (known then as Leopoldville). In the dense city slums, the virus spread quickly. Infected people traveled from the city along rivers and train lines to the other big cities of central Africa, such as Brazzaville, Lubumbashi, and Kisangani.

In the next few years, HIV-1 Group M left Africa. It first spread to Haiti, as Haitians who had been working in the Congo returned to their homeland after the country became independent from Belgium. Later, Haitian immigrants or American tourists may have brought HIV to the United States by the 1970s. That's about four decades after the virus became established in humans, and about one decade before five men in Los Angeles became sick with a strange form of pneumonia.

By the time scientists recognized HIV in 1983, in other words, the virus had already become a hidden global catastrophe. And by the time scientists began trying to fight the virus, HIV already had a huge head start. The annual death toll climbed through the 1980s and 1990s. Some scientists predicted the virus could be quickly stopped with a vaccine, but a series of failed experiments dashed their hopes.

It took years of hard work to stem the tide of HIV. Public

health workers found that they could dial down the transmission of the virus with social policies, such as controlling the use of needles and distributing condoms. Later, the invention of powerful anti-HIV drugs helped the fight enormously. Today, millions of people take a cocktail of drugs that interfere with the ability of HIV to infect immune cells and use them to replicate. In affluent countries such as the United States, these drug therapies have allowed some people to enjoy a long and relatively healthy life. As governments and private organizations deliver those drugs to poorer countries, victims of HIV are living longer there as well. In 2005, the annual death rate from HIV peaked at 2.5 million a year. Since then, it's slowly declined. In 2019, HIV killed 690,000 people.

In theory, the world could drive that number to zero, and the best hope for doing so would be to invent an HIV vaccine. Some scientists started work on a vaccine shortly after the discovery of HIV itself, but the result has been a string of huge disappointments. Vaccines that looked promising in experiments on cells failed to yield results in animals. The vaccines that protected animals didn't help people.

One reason for this failure has been that HIV mutates quickly, even for a virus. A century of replication inside millions of people has generated a vast amount of genetic diversity in HIV. A vaccine that confers protection against one version of HIV is often useless against another. It may be possible in the years to come to invent a vaccine that can work against them all. But our ignorance of the virosphere has given HIV a deadly head start.

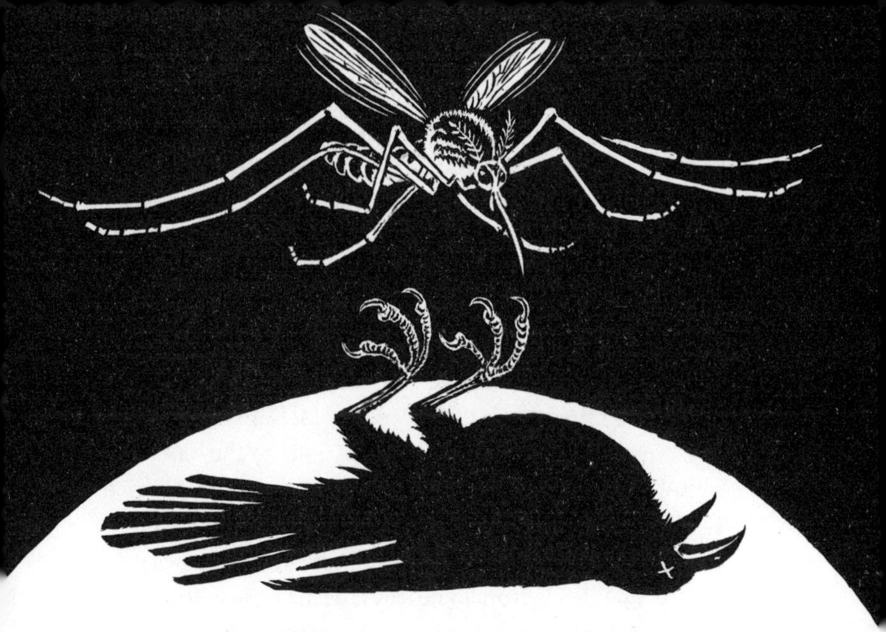

Becoming an American

THE GLOBALIZATION OF WEST NILE VIRUS

In the summer of 1999, the crows started to die.

Tracey McNamara was at the Bronx Zoo, where she works as the chief pathologist, when she noticed the dead crows lying on the ground. From her experience with death and disease in animals, she knew this was a bad sign. She feared that some new, deadly virus was sweeping through wild populations of birds around New York City. If the crows were catching the virus, they might pass it on to the zoo's collection of birds.

Over Labor Day weekend, her worst fears were realized. Three Chilean flamingoes suddenly died. So did a Himalayan Impeyan pheasant, a bald eagle, and a cormorant. The outbreak went on to claim more victims at the zoo, including black-billed magpies, a black-crowned night heron, laughing gulls, a Blyth's tragopan, bronze-winged ducks, and a snowy owl.

As zoo workers brought McNamara more dead birds, she examined their corpses to find a common thread that joined their deaths. They all showed the same signs of an infection that caused their brains to bleed. But McNamara could not figure out what pathogen was responsible, so she sent tissue samples to government laboratories. The government scientists ran test after test for the various pathogens that might be responsible. For weeks, the tests kept coming up negative.

Meanwhile, doctors in nearby Queens were getting worried, too. Among their patients, they were seeing a surge of encephalitis, an inflammation of the brain. The entire city of New York normally sees only about nine cases a year, but in August 1999, doctors in Queens found eight cases in one weekend. As the summer waned, more cases came to light. Some patients suffered fevers so dire that they became paralyzed, and by September nine had died. Initial tests pointed to a viral disease called Saint Louis encephalitis. But later tests failed to match the early results.

As the doctors struggled to make sense of the human outbreak, McNamara finally found the answer to her own mystery. The National Veterinary Services Laboratory in Iowa managed to grow viruses from the tissue samples she had sent them from the zoo's dead birds. They bore a resemblance to the Saint Louis encephalitis virus. McNamara wondered

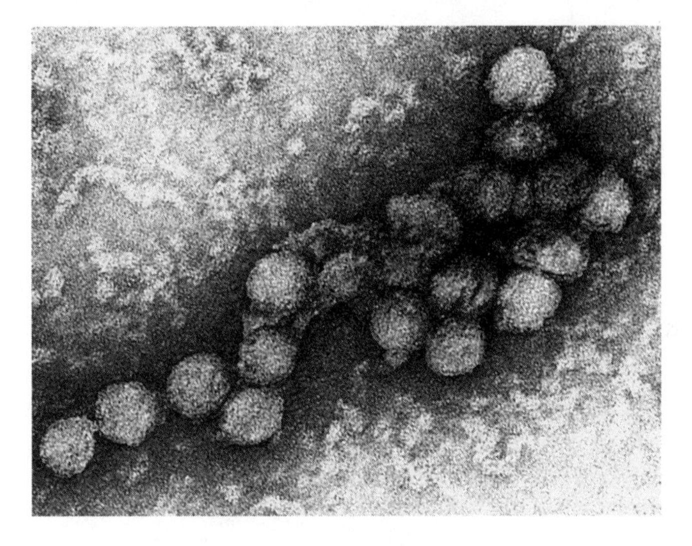

West Nile viruses in suspension

now if both humans and birds were succumbing to the same pathogen. But she'd have to know more about the viruses than looks alone.

McNamara convinced the Centers for Disease Control and Prevention to analyze the genetic material in the viruses. On September 22, the CDC researchers were stunned to find that the birds were not infected with the Saint Louis encephalitis virus, but an exotic pathogen called West Nile virus. Discovered in Uganda in 1937, West Nile virus was known to infect birds as well as people in parts of Asia, Europe, and Africa. But McNamara and the other researchers had not expected to find it in the Bronx. After all, it had never been seen in the Western Hemisphere before.

Meanwhile, the public health workers puzzling over New York's cases of encephalitis decided it was time to broaden their search as well. Two teams—one at the CDC and another led by Ian Lipkin, who was then at the University of California, Irvine—isolated the genetic material from the human viruses. They turned out to be West Nile viruses, too. No human in North or South America had ever suffered from it before.

The United States is home to many viruses that make people sick. Some are old and some are new. When the first humans made their way into the Western Hemisphere some 15,000 years ago, they brought papillomaviruses and a number of other viruses with them. In the sixteenth century, Europeans brought a fresh wave of infection to the Americas. New viruses, such as influenza and smallpox, killed millions of Native Americans. In later centuries, still more viruses arrived. HIV came to the United States in the 1970s, and at the end of the twentieth century, West Nile virus became one of America's newest immigrants. Since its arrival, it has settled comfortably in its new home. In its first 20 years in the United States, West Nile virus has spread to almost every state, infected an estimated seven million people, caused 2,300 deaths, and shows every sign of thriving for years to come.

West Nile virus does not spread in droplets through the air like influenza, or in bodily fluids like HIV. Instead, it is delivered in the bites of mosquitoes. When a mosquito lands on a person, it jams its syringe-like mouth into the skin. To prepare to slurp up blood, it first injects enzymes from its salivary glands into the wound. If the mosquito is infected with West Nile virus, it will inject some of these pathogens into the skin as well.

Once West Nile virus gets inside a human host, it drifts through the skin until it encounters an immune cell. In most people, this encounter spells a quick end to the infection. About 80 percent of people who get West Nile never feel sick. Yet even without experiencing symptoms, they produce potent antibodies that will keep them from ever getting infected again.

For the remaining 20 percent, an infection with West Nile virus doesn't clear so quickly. The immune cells that are supposed to wipe out the virus in the skin instead become infected with it. Some of them crawl to a lymph node, where the virus can then jump from cell to cell. The infected cells then stream out of the node and spread throughout the body. People who suffer a serious infection of West Nile virus may run a fever, suffer headaches, feel exhausted, or vomit. These symptoms usually pass once the immune system finally catches up with the infection, but in about 1 percent of infected people—most of them over the age of 50—the virus manages to reach the brain. It can infect neurons and kill them, and it can wreak even more havoc by triggering the immune system to produce a wave of inflammation.

For all the damage that West Nile virus can cause in some people's bodies, humans are not important to its long-term survival. Even our worst infections don't produce enough new viruses to infect a mosquito that bites us. We are dead ends for West Nile virus, in other words, as are dogs, horses, squirrels, and a number of other mammal species. In a bird, by contrast, West Nile virus can multiply into the billions within a few days of a mosquito bite.

To reconstruct the history of West Nile virus, scientists

have analyzed its genes much as they've done with other viruses like HIV. This research suggests that it first evolved in birds in Africa. Birds later carried it to other continents on their migrations, where it could infect new species. Along the way, West Nile virus also infected humans. In just one 1996 epidemic in Romania, 90,000 people came down with West Nile, leading to 17 deaths. Eventually the people in these regions developed immunity to the virus. The explosive outbreaks were replaced by lower, steadier infections.

It is striking that the United States was spared West Nile virus for so long. The genetic variations in West Nile virus found across the country suggest that it first arrived in 1998 and then went undetected for months before flaring up in New York. All the American strains of West Nile virus most closely resemble a sample isolated from a dead goose in Israel in 1998. Some scientists have speculated that a pet smuggler brought an infected bird from the Near East to New York. Others have wondered whether a virus-laden mosquito stowed away on a flight.

Whichever animal carried West Nile virus to the United States, it found an abundance of new hosts in which it could thrive. West Nile virus has been found in 62 species of mosquito native to the United States and 300 of its species of birds. A few birds in particular, including robins and sparrows, turned out to be particularly good incubators. Hopping from bird to mosquito to bird, West Nile virus spread across the entire United States in just four years. And from the United States, it soon spread north to Canada, and south to Brazil and Colombia.

Once West Nile virus arrived in the Western Hemisphere,

it settled into a regular cycle. In the spring, birds produce new generations of chicks that are helpless targets for virus-carrying mosquitoes. The percentage of infected birds goes up through the summer, and many mosquitoes get infected by feeding on them. Those mosquitoes can then bite people who are spending more time outdoors in the warm months of the year, giving them West Nile.

When the temperature drops in the fall, mosquitoes die across much of the United States, and the viruses can no longer spread. No one knows for sure how the virus survives the winter. It's possible that the virus population endures in the south, where the weather isn't so harsh on their insect hosts. It's also possible that when mosquitoes lay their eggs, they infect their offspring with West Nile. After overwintering, the infected eggs hatch in the spring, producing a new generation ready to spread the virus to more birds.

The life cycle of West Nile virus has made it particularly hard to fight. Measures that can drive down other viruses are useless against it. Washing hands and shutting down schools can help slow an influenza outbreak because the only way flu viruses travel to new hosts is inside tiny droplets released from sick people's mouths and noses. West Nile virus, by contrast, is actively delivered to new hosts by hungry mosquitoes. Some communities have tried to fight West Nile by spraying pesticides in mosquito breeding grounds, but these efforts have not entirely eradicated the insects and have threatened environmental harm.

West Nile virus is also hard to fight because we are its dead end. Many species of viruses, such as human papillomavirus and smallpox, are exquisitely adapted to our species and cannot survive in any other. But West Nile virus thrives in many species of birds. Even if doctors could somehow get rid of every West Nile virus infecting a human host, billions of birds would be generating a new batch for mosquitoes to deliver to us.

Unfortunately for those who get sick, there are still no antiviral drugs that can clear a West Nile infection. Nor is there a vaccine approved for use on people. When West Nile virus first arrived in the United States, several vaccine makers launched trials. They got as far as showing that some vaccines are safe and can make antibodies to West Nile virus. But the cost and demands required to run large-scale trials proved too much to justify the potential benefit. Horses have had better luck: veterinarians can give them an effective vaccine. Even endangered birds such as the California condor have gotten shots to protect them against the virus. We humans, it seems, will have to wait.

The story of West Nile virus has replayed itself twice in the years since. In 2013, a new mosquito-borne virus called chikungunya arrived in the Caribbean. It had first been identified during an outbreak in Tanzania in 1952. Its name means "to become contorted" in the Kimakonde language of southern Tanzania—a description of how its victims stoop with join pain. Nobody can say how chikungunya virus arrived in the Americas—whether it was a virus-infected traveler or a mosquito released from an airplane. The one clue scientists do have is the genetic material of the viruses. The Caribbean strain of chikungunya is nearly identical to a strain that has circulated in China and the Philippines. Somehow the virus leaped across the planet. And once it made the leap, it exploded. In the first year alone, chikungunya caused over a million infections in its new home.

Two years later, a new virus came to light in Brazil. Doctors became aware of it because hundreds of babies were born with drastically underdeveloped brains. It turned out that their mothers had been infected by yet another obscure mosquito-borne virus called Zika. It got its name from the Zika forest in Uganda, where it was discovered in a monkey in 1947. The following year, scientists discovered it in a mosquito in the same forest. In the decades that followed, Zika sporadically gave people fevers in East Africa until it caused its first large outbreak—not in Uganda, but thousands of miles away on Yap Island in the Pacific. Zika subsequently spread to many other countries in Asia; a 2014 study on Indonesian children found that 9 percent of them had antibodies to the virus.

In 2015, Zika finally arrived in the Americas. After ravaging Brazil, it was spread by mosquitoes and sexual contact into countries like Colombia and Mexico. The first cases in the United States turned up in the spring of 2016. Puzzlingly, the Zika virus didn't seem to create as much of a risk for birth defects as it spread north. By 2017, the Zika epidemic was burning itself out. Researchers aren't entirely sure why, but it appears that many people got infected without even knowing it and developed immunity. After the epidemic came to an end, the Zika virus didn't disappear, however. Thousands of people in South America continued to get sick each year from Zika, and scientists fully expect it to flare up again when the conditions are right.

The future looks rosy for West Nile virus and the other mosquito-borne viruses that have followed it to the Americas. That's because the future is going to be warm. Studies on West Nile across the United States over the past two decades have shown that it thrives in years in which the temperature is high. In places that also get enough rain, mosquitoes can reproduce faster and breed for more of the year. They also help the virus itself multiply faster inside the insects. Carbon dioxide and other heat-trapping gases are raising the average temperature in the United States, and climate scientists project it will continue to rise much higher in decades to come, with some regions expecting more humid, stormy conditions. The conditions may not only foster the growth of mosquitoes and viruses; they may produce warm winters that allow mosquitoes to push their ranges further north. Now that West Nile virus has made a new home, we're making that home more comfortable.

The Pandemic Age

WHY COVID-19 SHOULD HAVE COME AS NO SURPRISE

Li Wenliang worked as an ophthalmologist in a hospital in Wuhan, a sprawling city in eastern China with a population of 11 million. In December 2019, the 34-year-old doctor learned that his hospital was quarantining seven patients suffering from severe pneumonia. They all worked at the same whole-sale fish market, suggesting that an outbreak was underway in the city. Local authorities were silent about the pneumonia, but Li was alarmed by what he learned.

The symptoms of the seven patients—fevers, hacking coughs, fluid-filled lungs—reminded him of a disease that swept through China 17 years earlier. Severe Acute Respiratory Syndrome, or SARS for short, had been caused by a type of virus called a coronavirus. While most coronaviruses in humans caused mild colds, SARS killed 10 percent of its victims. Fortunately, quarantines brought the SARS outbreak to a stop, and the virus had never been seen since.

Now Wuhan was seeing a cluster of SARS-like cases. A fellow doctor at the hospital showed Li a test result from one of the patients, revealing the presence of a coronavirus. On social media, Li belonged to a closed group of doctors who were all alumni of Wuhan University. He posted a warning to his friends on December 30, urging them and their families to be on the alert.

Someone took a screenshot of the message, and soon it went viral online. Rumors had been circulating for days about the pneumonia, and here was confirmation from a doctor in a major hospital—the first time that a medical professional raised the alarm.

"I only wanted to remind my university classmates to be careful," he later told a reporter for CNN. "When I saw them circulating online, I realized that it was out of my control and I would probably be punished."

He was right. Officials at Li's hospital summoned him to explain how he knew about the cases, and then, on January 3, he was called to the local police office to sign a statement that he "severely disturbed public order" by spreading claims "that were not factual and broke the law." Li promised not to commit any further unlawful acts.

By then, China was sharing the same information to the World Health Organization it punished Li for spreading. With each day, more cases of pneumonia showed up across Wuhan. Li went back to work, trying to avoid further trouble. A few days later, he treated a patient for glaucoma. Other than her eye condition, she seemed healthy and Li didn't take any precautions for the appointment. Later the patient became sick and infected her whole family. On January 10, Li began to cough. "I got careless," he told the *New York Times*.

Soon Li was struggling to breathe on his own. He was admitted to the hospital, where his fellow doctors gave him a mask to supply him with oxygen. Because Li's pneumonia was caused by a virus instead of bacteria, antibiotics were of no help. All his doctors could do was wait and hope that he would recover—a reasonable hope, given that he was a healthy, 34-year-old man. Because the virus was so infectious, he had to be kept in strict isolation. His pregnant wife and four-year-old child could speak to him only by video. When Li spoke with the *Times* at the end of January, he said he was confident that he would eventually get out of his hospital bed. "It will take me another 15 days or so to recover," he said. "I will join medical workers in fighting the epidemic. That's where my responsibilities lie."

A week later, Li was dead. Soon the virus had an official name—SARS-CoV-2—and the disease it caused was dubbed COVID-19. By the time his wife gave birth in June, SARS-CoV-2 had spread across the planet. Nearly eight million people had tested positive for the virus, and it was likely that tens of millions more had become infected. The official death count had passed 430,000, although the true toll was likely much higher. As the virus took hold in one country after another, they had little choice but to lock down their citizens to slow its onslaught. The result was the greatest economic downturn since the Great Depression, wiping out trillions of dollars from the world's economy and pushing hundreds of millions of people into poverty.

It's possible that if doctors like Li had been able to sound the alarm sooner, much of this global catastrophe might have been avoided. We may never know the identity of the first person to become infected with COVID-19—the Patient Zero of this new pandemic. But we should honor the memory of its first hero.

COVID-19 took many people by surprise, but it shouldn't have. Virologists had been warning of the threat of emerging viruses for decades. "The worldwide epidemic of the acquired immunodeficiency syndrome (AIDS) demonstrates that infectious diseases are not a vestige of our premodern past, but, like disease in general, the price we pay for living in the organic world," the virologist Stephen Morse wrote in 1991.

Morse issued this warning as HIV was emerging as a global threat after evolving from a chimpanzee virus. Morse and other virologists worried that other animal viruses might spill over the species barrier as well. Their names—Rift Valley fever, monkeypox, Ebola—were only familiar in the 1990s to virologists. Every now and then they infected a few people with horrible results before disappearing back into their animal hosts. But one of them, under the right conditions, might become the next HIV, the next pandemic flu. And the more pressure humans exerted on the habitats where these animals lived, by logging rain forests and clearing wilderness for large-scale farming, the more likely that a virus would emerge.

Morse warned that the next threat might come from a virus that didn't even have a name yet. "Unless concerted efforts are made to search, viruses tend to be discovered either by happenstance," he wrote, "or, as has been the case for almost all human viral diseases, when they finally do emerge to cause dramatic disease outbreaks somewhere in the Western world."

SARS emerged 11 years after Morse issued this warning. At first, it seemed to fit his prophecy to a T. In November 2002, a Chinese farmer came to a hospital suffering from a high fever and died soon afterward. Other people from the same region of China began to develop the disease as well, but it didn't reach the world's attention until an American businessman flying back from China developed a fever on a flight to Singapore. The flight stopped in Hanoi, where the businessman died. Soon, people were falling ill in other countries as far away as Canada.

Scientists began searching samples from SARS victims for a cause of the disease. Malik Peiris of the University of Hong Kong led the team of researchers who found it. In a study of fifty patients with SARS, they discovered the same viruses growing in two of them. Peiris and his colleagues sequenced the genetic material in the new virus and then searched for matching genes in the other patients. They found a match in 45 of them.

They were coronaviruses—a type of virus that got its name from a crown of spike-shaped proteins on its surface. The scientists who first studied the viruses in the 1960s were reminded of the corona that becomes visible around the sun during an eclipse. Since human coronaviruses known up till then caused only mild colds, the discovery of one that could cause fatal pneumonia was surprising. When Morse and his colleagues came up with a list of the kinds of viruses to worry about, coronaviruses did not make the cut.

Nevertheless, SARS arose precisely as Morse and his colleagues had prophesied. To trace the origin of the virus—known as SARS-CoV—scientists used the same approach they had used to determine the origin of HIV: they drew a family tree of the virus and then looked in animals for close relatives. They found that SARS-CoV arose not in primates but in Chinese horseshoe bats.

It's possible that SARS-CoV first spread to another animal before attacking humans. Scientists found the virus in a catlike mammal called a civet that was a common sight in Chinese animal markets. Yet it's also possible that SARS-CoV jumped directly from bats to humans. Someone may have eaten an infected bat or become contaminated by coming in contact with its droppings. Whatever its route to our species, the virus turned out to have the right biology for spreading from person to person.

Fortunately, it turned out that people with SARS became infectious only after they started showing the symptoms such as a fever and cough. As soon as they became sick, doctors could swiftly put them in isolation and prevent them from spreading the virus. All told, SARS caused about 8,000 cases and 900 deaths before it disappeared. Compared to an ordinary year of flu infections, the emergence of SARS was a dodged bullet. But if SARS had come from Chinese horseshoe bats, scientists knew, it could come again.

A decade later, a new coronavirus appeared in Saudi Arabia. In 2012, doctors at Saudi hospitals noticed that some of their patients were becoming ill with a respiratory disease they couldn't identify. It was like SARS, but more lethal, killing nearly a third of the patients. The disease came to be known as MERS, short for Middle Eastern Respiratory Syndrome, and soon after, virologists isolated the coronavirus that caused it. It was a fairly close cousin to SARS-CoV. The closest relatives to MERS-CoV could also be found in bats—but in the case of MERS, the bats lived in Africa, not China.

How African bats could have caused a Middle Eastern epidemic was a question without an obvious answer. But an important new clue emerged when scientists examined the mammals that many people in the Middle East depend on for their survival: camels. They began to find camels rife with MERS viruses, oozing from their noses in drops of mucous. One possible explanation for the origin of MERS is that bats passed on the virus to camels in North Africa. There's a healthy trade in camels from North Africa to the Middle East. A sick camel may have carried the virus to its new home.

As scientists reconstructed the history of MERS, there was good reason to fear a global outbreak even worse than SARS. Each year, over two million Muslims travel to Saudi Arabia for the annual pilgrimage known as the hajj. It was easy enough to imagine the MERS virus spreading swiftly among the crowds, and then traveling with the pilgrims to their homes around the world. But so far, that hasn't happened. Every few months, a new outburst of a few dozen cases of MERS has flared. The virus has struck 27 countries, causing a total of 2,562 cases as of November 2020 and claiming 881 lives. Most

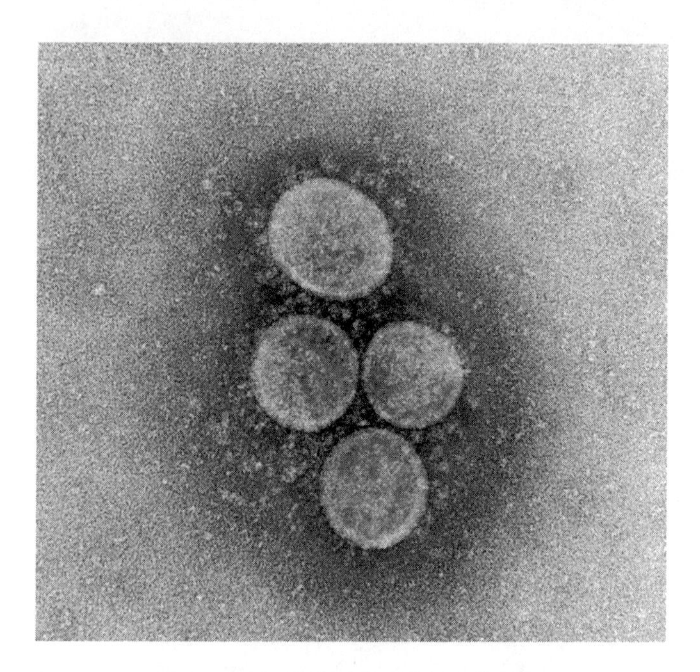

COVID-19 virus particles, isolated from a patient

of the outbreaks have occurred in hospitals, leading scientists to suspect that MERS can successfully invade only people with weakened immune systems.

As frightening as SARS and MERS were, the world responded with complacency. The viruses didn't trigger global concern because they couldn't manage to reach very many people. That was not the case for the coronavirus that causes COVID-19.

The genetic makeup of SARS-CoV-2 shows that it is closely related to SARS-CoV. It's possible that these coronaviruses descend from a common ancestor that infected bats several hundred years ago. For centuries, their ancestors circulated in bats across China. They mutated and adapted to their airborne hosts. They also mixed their genes together into new combinations when two coronaviruses infected a single animal. Of all the coronaviruses that scientists have found in Chinese bats, the ones most closely related to SARS-CoV-2 diverged from it decades ago. The immediate origins of the COVID-19 pandemic are therefore shrouded in mystery for now.

It's likely that SARS-CoV-2 took the same basic path as its coronavirus cousins, not to mention other viruses like HIV and influenza. In 2019, a single person in China came down with a coronavirus infection. That first human host might have been a Chinese farmer far from Wuhan. The virus may have already been well adapted to infect our airways and then gradually adapted to moving from human to human instead of bat to bat. Once it reached the city of Wuhan, the virus encountered millions of people living and working in close quarters, where a single infected person could infect dozens of other people.

In some ways, SARS-CoV-2 worked identically to SARS-CoV. The two coronaviruses used the same protein on the surface of airway cells, called ACE2, to invade. They both could trigger the immune system to produce an overwhelming, destructive response that devastated a patient's lungs. But the new coronavirus was different in some crucial ways. It was far less lethal for one thing. Roughly 1 in 200 of people who get COVID-19 will die, as opposed to the 1 in 10 claimed by SARS. But unlike people with SARS, people who get COVID-19 can spread the virus days before they start showing symptoms. In a fifth of COVID-19 cases, people develop no symptoms whatsoever. As a result, COVID-19 spread across China and into other countries long before public health authorities realized that they had a disaster on their hands. And once the virus established itself in a new country, it was often impossible to rein it in with the strategies that worked so effectively on SARS 17 years earlier.

Many people would not realize that they even had COVID-19 for months after their infections. They would need an antibody test to show that their immune systems had fought off the virus. Some people were fortunate to suffer few symptoms, but many people were laid low for days or weeks. About 20 percent of people who were infected had to be hospitalized. Doctors encountering COVID-19 for the first time found it dramatically different than influenza or other respiratory diseases they were familiar with. It could ravage people's lungs, and the virus could also spread to other parts of the body, to cause heart attacks, kidney failure, and strokes from blood clots.

The pandemic took different courses in different countries, largely as a result of how seriously their governments prepared for the worst. South Korea, for example, had suffered badly in the SARS epidemic and then endured a brutal outbreak of MERS in its hospitals. The government recognized that coronaviruses might well deliver another attack. They stored up supplies of protective gear for their healthcare workers and invested in public health experts who could trace viruses as they spread from person to person. On January 20, 2020, South Korea confirmed its first case of COVID-19 and immediately took decisive action. They developed a genetic test to detect infections of COVID-19. To make it easy for people to get safely tested, they invented drive-through testing, where health-care workers in protective gear could lean into cars and run swabs into people's noses. When they discovered that a church had become a hot spot of infection, the South Korean government dispatched a squadron of contact tracers to find everyone who had been exposed. By the end of 2020, they had only 60,000 cases and 900 deaths.

The United States saw its first case on the same day, but it suffered far more carnage. The government chose not to use an existing test for SARS-CoV-2, instead creating its own. Through staggering bureaucratic incompetence, it took weeks to discover that the test was faulty. Cutting-edge biology labs at American universities could have easily produced a working test, but the government blocked any attempts. Through the rest of January and on through February, the United States did barely any testing, focusing most of its attention on travelers from China. The Trump administration put a China travel ban in place, but it was a largely useless move, since the virus had already spread to a number of other countries. New York, it would later turn out, received most of its incoming virus from Europe. In March, hospitals filled with COVID-19 cases. Unlike their Korean counterparts, American health-care workers often had to struggle to find enough protective equipment to shield themselves from the highly contagious virus. Some wore garbage bags. Cities across the country—Los Angeles, Seattle, Chicago, Detroit, and more—all became Wuhans of their own. New York City suffered over 25,000 deaths by the end of the year, well over 1,000 times more lost lives than in Seoul, a city of about the same size. Across the entire country, over 18.7 million Americans tested positive for COVID-19 by the end of 2020 and some 350,000 were dead.

By the end of the year, however, people could see beyond the despair, because vaccines were coming. The search for COVID-19 vaccines had begun in January 2020, as soon as Chinese scientists isolated SARS-CoV-2 and sequenced its genome. Some researchers used traditional methods, dousing the coronavirus in chemicals to inactivate it, much as Jonas Salk had done in the 1950s to invent the polio vaccine. Others used newer methods, such as creating molecules of RNA that instructed people's own cells to make viral proteins. In November, clinical trials began demonstrating that some of the vaccines were able to protect volunteers from COVID-19. And in December, mass vaccination campaigns began around the world. It typically takes a decade or more to bring a new vaccine through testing and into the clinic. Before COVID-19, the record for a vaccine had been four years, for mumps. Scientists shattered that record in order to start bringing the pandemic to an end.

For the sake of humanity, we must learn from this experience. There will be more COVIDs—perhaps a COVID-24, a COVID-31, a COVID-33. Coronaviruses are only one group of viruses that researchers know can produce new human diseases. And virologists are keenly aware that they have only begun to explore the diversity of the virosphere. To reduce that ignorance, scientists are surveying animals, searching for bits of genetic material from viruses. But because we live on a planet of viruses, that task is enormous. Ian Lipkin and his colleagues at Columbia University trapped 133 rats in New York City and discovered 18 new species of viruses that are closely related to human pathogens. In another study in Bangladesh, they examined a bat called the Indian flying fox and tried to identify every single virus that calls it home. They identified 55 species, 50 of which were new to science.

We can't say which, if any, of these newly discovered viruses will create a new pandemic. But that doesn't mean that we can simply ignore them. Instead, we need to stay vigilant, so that we can block them before they get a chance to make the great leap into our species.

The Long Goodbye

THE DELAYED OBLIVION OF SMALLPOX

As I write in 2021, the ultimate fate of COVID-19 remains uncertain. Will the coronavirus keep grinding through the world's populations, killing millions? Will vaccines offer enough protection to drive down the virus to low levels, with antivirals at the ready to render COVID-19 a mild illness? Will it find refuge in a cave full of bats, ready to spill back over someday like SARS? Or might we eradicate it altogether?

That last possibility is the least likely, if history is any guide. Medicine has managed to completely eradicate only a single species of human virus from nature. The distinction goes to the virus that causes smallpox. But what a virus to wipe out. Over the past few thousand years, smallpox may have killed more people than any other disease on Earth.

The origins of smallpox remain murky, but by the fourth century, physicians in China were leaving carefully observed details of the course of the disease. The virus spreads through the air, and a week after the infection people start to experience chills, a blazing fever, and agonizing aches. The fever ebbs after a few days, although the virus is far from done with its victim. Red spots develop inside the mouth, then on the face, and then over the rest of the body. The spots fill with pus and cause stabbing pain. About a third of people who get smallpox eventually die. In the survivors, scabs cover over the pustules, which leave behind deep, permanent scars.

The oldest direct evidence of smallpox comes from Viking skeletons dating back to the seventh century. Their bones still hold on to fragments of the virus's genes. In the centuries that followed, the virus turned up in new places and wreaked havoc. When it arrived in Iceland in 1241, it promptly killed 20,000 of the island's 70,000 inhabitants. Between 1400 and 1800, smallpox killed an estimated 500 million people every century in Europe alone. Its victims included sovereigns such as Czar Peter II of Russia, Queen Mary II of England, and Emperor Joseph I of Austria.

It was not until Columbus's arrival in the Caribbean that the people of the Americas got their first exposure to the virus. The Europeans unwittingly brought a biological weapon with them that gave the invaders a brutal advan-

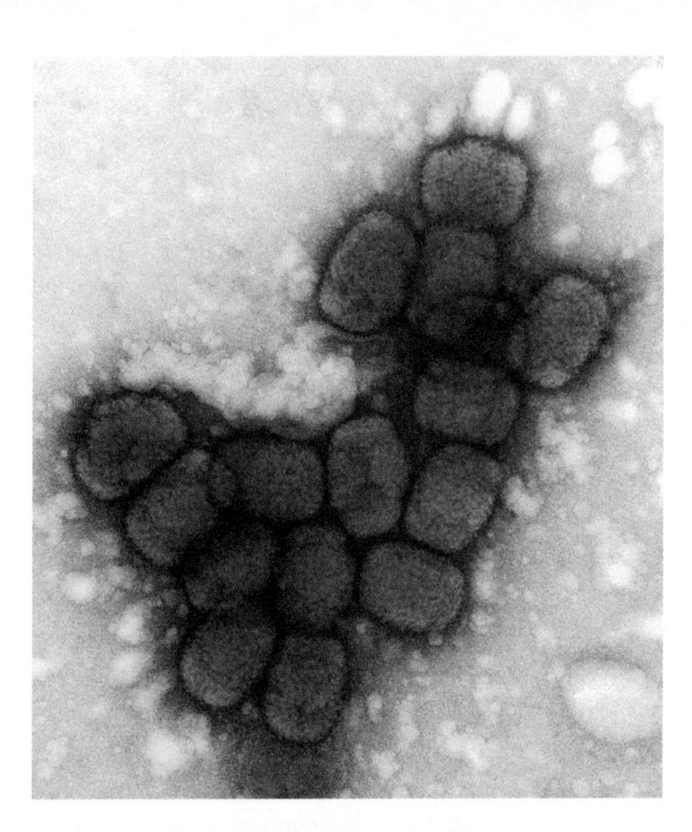

Smallpox viruses in suspension

tage over their opponents. With no immunity whatsoever to smallpox, Native Americans died in droves when they were exposed to the virus. In Central America, over 90 percent of the native population is believed to have died of smallpox in the decades following the arrival of the Spanish conquistadores in the early 1500s.

The first effective way to prevent the spread of smallpox probably arose in China around the year 900. Physicians would rub a scab from a smallpox victim into a scratch in the skin of a healthy person. (Sometimes they administered it as an inhaled powder instead.) Variolation, as this process came to be called, typically caused just a single pustule to form on the inoculated arm. Once the pustule scabbed over, a variolated person became immune to smallpox.

At least, that was the idea. Fairly often, variolation would trigger more pustules, and in 2 percent of cases, people died. Still, a 2 percent risk was more attractive than the 30 percent risk of dying from a full-blown case of smallpox. Variolation spread across Asia, moving west along trade routes until the practice came to Constantinople in the 1600s. As news of its success traveled into Europe, physicians there began to practice variolation as well. The practice triggered religious objections that only God should decide who survived the dreaded smallpox. To counteract these suspicions, doctors organized public experiments. Zabdiel Boylston, a Boston doctor, publicly variolated hundreds of people in 1721 during a smallpox epidemic; those who had been variolated survived the epidemic in greater numbers than those who had not been part of the trial. During the Revolutionary War, George Washington ordered that all his soldiers be variolated, to spare his army what he called "the greatest of all calamities that can befall it."

No one at the time knew why variolation worked, because nobody knew what viruses were or how our immune systems fought them. The treatment of smallpox moved forward mainly by trial and error. In the late 1700s, the British physician Edward Jenner invented a safer smallpox vaccine based on stories he heard about how milkmaids never got the disease. Cows can get infected with cowpox, a close relative of smallpox, and so Jenner wondered if it provided some protection. He took pus from the hand of a milkmaid named Sarah Nelmes and inoculated it into the arm of a boy. The boy developed a few small pustules, but otherwise he suffered no symptoms. Six weeks later, Jenner variolated the boy—in other words, he exposed the boy to smallpox, rather than cowpox. The boy developed no pustules at all.

Jenner introduced the world to this new, safer way to prevent smallpox in a pamphlet he published in 1798. He dubbed it "vaccination," after the Latin name of cowpox, *Variolae vaccinae*. Within three years, over 100,000 people in England had gotten vaccinated against smallpox, and vaccinations spread around the world. In later years, other scientists borrowed Jenner's techniques and invented vaccines for other viruses. From rumors about milkmaids came a medical revolution.

As vaccines grew popular, doctors struggled to keep up with the demand. At first they would pick off the scabs that formed on vaccinated arms and use them to vaccinate others in turn. But since cowpox occurred naturally only in Europe, people in other parts of the world could not easily acquire the virus as Jenner had. In 1803, King Carlos of Spain came up with a radical solution: a vaccine expedition to the Americas and Asia. Twenty orphans boarded a ship in Spain. One of the orphans had been vaccinated before the ship set sail. After eight days, the orphan developed pustules, and then scabs. Those scabs were used to vaccinate another orphan, and so on through a chain of vaccination. As the ship stopped in port after port, the expedition delivered scabs to vaccinate the local population.

Physicians struggled throughout the 1800s to find a better way to deliver smallpox vaccines. Some turned calves into vaccine factories, infecting them repeatedly with cowpox. Along the way, the cowpox got mixed up with horsepox, a closely related virus. In the early 1900s, as the nature of viruses emerged, researchers left the calves behind and started making the vaccines in batches of cells. They could now make huge amounts of the stuff, with guaranteed purity. Countries ordered up enough vaccines to launch eradication campaigns. The work was slow and spotty, though; even in the twentieth century, smallpox is estimated to have killed 300 million people.

Finally, in the 1950s, the World Health Organization began contemplating the possibility that a concerted campaign could wipe smallpox off the face of the Earth. The advocates of an eradication campaign built their case on the biology of the virus. Unlike West Nile virus or influenza, smallpox infects only humans, not animals. If it could be systematically eliminated from every human population, there would be no need to worry that it was lurking in pigs or ducks, waiting to strike again. What's more, smallpox is an obvious disease. Unlike a virus like HIV, which can take years to make itself known, smallpox declares its gruesome presence in a matter of days. Public health workers would be able to identify outbreaks and precisely track them.

Yet the idea of eradicating smallpox still met with skepticism. If everything went exactly according to plan, an eradication project would require years of labor by thousands of trained workers, spread across much of the world, toiling in many remote, dangerous places. Public health workers had already tried to eradicate other diseases, like malaria, and failed. Why should smallpox be any easier to destroy?

The skeptics lost the debate, however, and in 1965, the World Health Organization launched the Intensified Smallpox Eradication Programme. Public health workers deployed a new prong-shaped needle that could deliver smallpox vaccine far more efficiently than regular syringes. The world's supply of smallpox vaccine could thus be stretched much further than before. Public health workers also realized they didn't have to reach the impossible goal of vaccinating every person on Earth. Instead, they only had to identify new smallpox outbreaks and take quick action to snuff them out. They quarantined victims and then vaccinated people in the surrounding villages and towns. The smallpox spread outward like a forest fire, but soon hit the firebreak of vaccination and died out. Outbreak by outbreak, the virus was beaten back, until the last case was recorded in Ethiopia in 1977. The world was now free of smallpox.

Ever since the campaign came to a close, it's served as proof that at least some pathogens can be wiped out. A few other campaigns have followed in its wake, but only one other virus has been eradicated successfully so far. For centuries, the rinderpest virus tormented dairy farmers and cattle herders by killing off entire herds of cows in deadly sweeps. Over the course of the 1900s, veterinarians carried out a series of vaccination campaigns against rinderpest, but they were never thorough enough to vanquish the virus, allowing it to bounce back again and again.

In the 1980s, rinderpest experts rethought their whole approach to the virus and began planning out a new campaign that would wipe it out for good. In 1990, vaccine developers created a cheap, stable rinderpest vaccine that could be transported on foot to even the most remote nomadic tribes. In 1994 the Food and Agricultural Organization used the vaccine to launch a global eradication program. They would gather information about sick cows from community workers and distribute vaccines where they were needed to keep infected animals from sickening healthy ones.

In country after country, rinderpest vanished. But wars would stop the campaigns and allow the virus to return to cleared territories. "Rinderpest is a prime candidate for eradication. Why has it not happened?" asked Sir Gordon Scott, a leader of the campaign, in a 1998 paper. "The major obstacle is 'man's inhumanity to man,'" he concluded. "Rinderpest thrives in a milieu of armed conflict and fleeing refugee masses."

Scott turned out to be too pessimistic. In 2001, just three years after he wrote his gloomy predictions, veterinarians recorded their last case of rinderpest: a wild buffalo in Mount Meru National Park in Kenya. The FAO waited another decade to see whether any other animals fell ill. None did, and in 2011, they announced rinderpest had been eradicated.

Other eradication campaigns have come tantalizingly close to victory, only to bog down in the endgame. Polio, for example, was once a worldwide threat, leaving millions of children paralyzed or trapped in iron lungs. Years of eradication efforts have eliminated the virus from much of the world. In 1988, 350,000 people got polio. In 2019, only 176 did. In 1988, polio was endemic in 125 countries. In 2019 it remained endemic in only Afghanistan and Pakistan. Yet those two countries have withstood years of efforts to eradicate the virus. Wars and poverty have gotten in the way of vaccination campaigns. Making matters worse, Taliban insurgents began to view vaccine campaigns as a threat and systematically assassinated vaccine workers. If polio is allowed to rebound, it could surge through Afghanistan and Pakistan and spread to neighboring countries, causing 200,000 cases a year by 2030.

As we begin eradicating viruses, we're also discovering that they can endure in all sorts of unnerving ways. In the late 1900s, as smallpox eradication workers traveled the world to wipe out the virus, scientists were also breeding it in their laboratories in order to study it. When the World Health Organization officially declared smallpox eradicated in 1980, those laboratory stocks remained. All it would take to reverse its eradication would be for someone to accidentally set the virus loose.

The World Health Organization decided that all the laboratory stocks would eventually have to be destroyed. But in the interim, they would still allow scientists to conduct research on the virus under strict rules. Only two approved laboratories could hold on to the remaining stocks of smallpox: one in the Siberian city of Novosibirsk in the Soviet Union, and one at the US Centers for Disease Control and Prevention in Atlanta, Georgia. Over the next three decades, smallpox research continued under the World Health Organization's watchful eye. Scientists learned how to engineer lab animals to become infected with smallpox, allowing them to better understand its biology. They analyzed its genome, worked on better vaccines, and found drugs that showed promise as cures for smallpox. And during that time, the WHO debated exactly when they should destroy the virus once and for all.

Some experts argued that there was no reason to wait. As long as smallpox existed—no matter how carefully controlled—the risk remained that the virus could escape and kill millions of people. Terrorists might even try to use it as a biological weapon. Raising the risk even higher, the world's immunity to smallpox was waning since no one was getting vaccinated against it any longer.

But other scientists urged holding on to smallpox stocks. They pointed out that the eradication campaign might not, in fact, have been a complete success. In the 1990s, Soviet defectors revealed that their government had set up labs to produce a weaponized form of smallpox, one that could be loaded in missiles and launched at enemy targets. After the fall of the Soviet Union, those biological warfare labs were abandoned. No one knows what ultimately happened to the smallpox viruses used for that research. We are left with the terrifying possibility that ex-Soviet virologists sold smallpox stocks to other governments or even terrorist organizations.

Opponents of smallpox eradication argued that the risk of new outbreaks—no matter how small—justifies more research on the virus. There's still so much we don't yet know about it. Smallpox can infect only a single species—humans—while all its relatives, called orthopox viruses, can infect several species. No one knows what makes smallpox so fussy. If a smallpox outbreak should occur in years to come, fast diagnosis could save untold lives. In order to develop cutting-edge tests, scientists will need to evaluate them to be sure that they distinguish between smallpox and other orthopox viruses. Only live smallpox viruses will do for such experiments. Likewise, scientists could use the viruses to develop better vaccines and antiviral drugs.

The argument over smallpox did not resolve itself in a clean decision—only an agreement to bring up the matter in the future. But as the disagreement dragged on for years, advances in technology changed the very terms of the debate.

In the 1970s, as public health workers were wiping out smallpox, geneticists developed the first methods for reading the sequence of genes. In 1976, they read all the genetic material—the genome—in a bacteriophage called MS2. It was the first genome to be fully sequenced. The choice of a virus for the first genome was no accident: scientists wanted to start small. While the human genome contains over 3 billion genetic "letters," MS2's genome has just 3,569—almost a million times fewer.

In the years that followed, scientists read the genomes of other viruses, including smallpox in 1993. By comparing its genome to those of other viruses, the scientists gained some clues to the workings of smallpox proteins. Researchers went on to sequence the genomes of smallpox strains from across the world, revealing that there was little variation between them—an important clue for researchers planning on preparations for future smallpox outbreaks.

The invention of genome-sequencing technology opened the way for another major advance: scientists began to assemble bases to synthesize genes from scratch. At first they assembled short stretches of genetic material. Even at that early stage, Eckerd Wimmer, a virologist at Stony Brook University, realized that viruses had genomes that were small enough that they could be synthesized in full. In 2002, he and his colleagues used the poliovirus genome as a guide for the creation of thousands of short DNA fragments. They then used enzymes to stitch the fragments together and used the DNA molecule as a template for making a corresponding RNA molecule—in other words, a physical copy of the entire poliovirus genome. When Wimmer and his colleagues added that RNA to test tubes full of bases and enzymes, live polioviruses spontaneously assembled. They had, in other words, made polio from scratch.

Wimmer argued that scientists could use this newfound power to help humanity. They could build viruses with precise changes to the genomes of viruses to better understand how they worked. They could rewrite virus genomes to create harmless versions of the most dangerous threats to human health, turning them into new vaccines. Wimmer went on to cofound a vaccine company that began using synthetic viruses as experimental vaccines for diseases including influenza, Zika, and COVID-19.

Still, critics worried that Wimmer's technology might fall into the wrong hands and that someone might start making viruses to unleash on the world. But they didn't worry much about synthetic smallpox at first. Smallpox has about 30 times more DNA than polio, making a synthetic version so difficult to assemble that the threat seemed more like science fiction.

It moved a lot closer to reality in 2018. David Evans, a

virologist at the University of Alberta, and his colleagues synthesized horsepox, one of the harmless cousins of smallpox. They took advantage of a set of powerful genetic tools that had been created in the years since Wimmer's work. The scientists emailed the sequences for 10 long pieces of horsepox DNA to a mail-order company, which synthesized the molecules and sent them to the scientists. Each segment was harmless on its own. But when Evans and his colleagues injected them all into a cell, it welded the pieces into a single DNA molecule. And that new molecule could generate viable horsepox viruses.

"The world just needs to accept the fact that you can do this," Evans told a reporter for *Science*. It cost him just $100,000 to do the job.

After thousands of years of suffering and puzzling over smallpox, we have finally come to understand it and halt its relentless destruction. And yet, by understanding smallpox, we have ensured that it can never be utterly eradicated as a threat to humans. The knowledge we have gained about viruses has given smallpox its own kind of immortality.

EPILOGUE

The Alien in the Water Cooler

GIANT VIRUSES AND WHAT IT MEANS TO BE A VIRUS

Wherever there is water on Earth, there is life. The water may be a Yellowstone geyser, a pool in the Cave of Crystals, or a cooling tower sitting on the roof of a hospital.

In 1992, a microbiologist named Timothy Rowbotham scooped up some water from a hospital cooling tower in the

English city of Bradford. He put it under a microscope and saw a welter of life. He saw amoebae and other single-celled protozoans, about the size of human cells. He saw bacteria, about 100 times smaller. Rowbotham was searching for the cause of an outbreak of pneumonia that had been raging through Bradford. Among the microbes he found in the cooling tower water, he thought he found a promising candidate: a sphere of bacterial size, sitting inside an amoeba. Rowbotham believed he had found a new bacterium and named it in honor of his city: *Bradfordcoccus*.

Rowbotham spent years trying to make sense of *Bradfordcoccus*, to see whether it was the culprit in the pneumonia outbreak. He tried to identify its genes by finding matches with genes in other species of bacteria. But he couldn't find any. In 1998, budget cuts forced him to close his lab down. Rather than destroy his puzzling *Bradfordcoccus*, he arranged for French colleagues to store his samples.

For years, *Bradfordcoccus* languished in obscurity, until Bernard La Scola of Mediterranean University decided to take another look at it. As soon as he put Rowbotham's samples under a microscope, he realized something was not right.

Bradfordcoccus did not have the smooth surface of spherical bacteria. Instead, it was more like a soccer ball, made up of many interlocking plates. And from those plates, La Scola saw hairlike threads of protein radiating in all directions. The only things in nature known to have these kinds of shells and threads were certain kinds of viruses. But La Scola knew, like all microbiologists at the time knew, that something the size of *Bradfordcoccus* could not be a virus, because it was 100 times too big.

And yet a virus is exactly what *Bradfordcoccus* turned out to be. When La Scola and his colleagues examined it further, they discovered that it reproduced by invading amoebae and forcing them to build new copies of itself. Only viruses reproduce this way. La Scola and his colleagues gave *Bradfordcoccus* a new name to reflect its viral nature. They called it a mimivirus, partly in honor of the virus's ability to mimic bacteria.

The French scientists set out to analyze the genes of the mimivirus. Rowbotham had tried—and failed—to match its genes to those of bacteria. The French scientists had better luck. The mimivirus turned out to have virus genes—and a lot of them. Before the discovery of mimiviruses, scientists had become accustomed to finding only a few genes in a virus. But mimiviruses have 1,018 genes. It was as if someone took the genomes of the flu, the cold, smallpox, and 100 other viruses and stuffed them all inside one protein shell. The mimivirus even had more genes than some species of bacteria. In both its size and its genes, mimivirus had broken cardinal rules for being a virus.

La Scola and his colleagues published their first report about the remarkable mimivirus in 2003. They wondered if it was unique. Perhaps there were other giant viruses also hidden in plain sight. They collected water from cooling towers in France and added amoebae to it, to see whether anything in the water might infect them. Soon the amoebae were exploding, releasing giant viruses.

These were not mimiviruses, however. They were another species, with 1,059 genes, setting a new record for the biggest genome in a virus. While the new virus looked a lot like mimivirus, its genome was profoundly different. When the

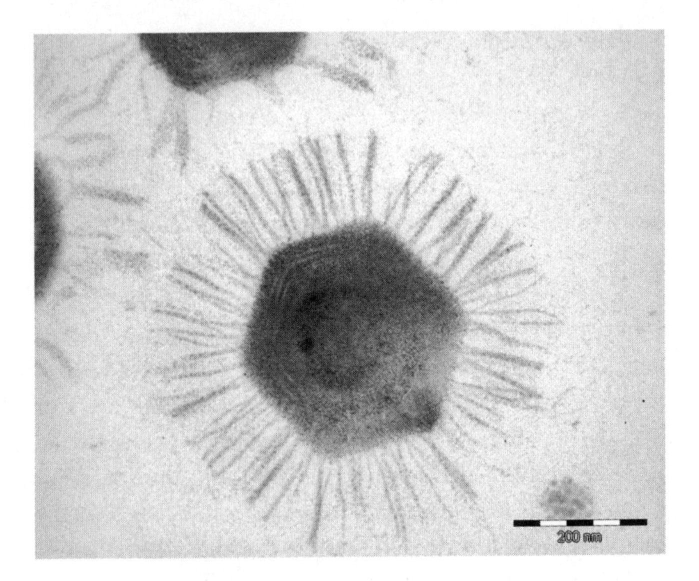

Mimivirus, one of the largest known viruses

scientists lined up the new virus's genes with those of mimivirus, they could match only 833 of them. The other 226 were unique. Other researchers joined the hunt, and they began finding giant viruses everywhere: in rivers, in oceans, in lakes buried under Antarctic ice. In the sea floor off the coast of Chile, scientists found giant viruses with 2,556 genes—for now, the biggest viral genome by far.

Scientists have even begun finding giant viruses lurking in animals. La Scola and his colleagues collaborated with Brazilian scientists to study serum samples taken from mammals. They found antibodies to giant viruses in monkeys and cows. The researchers have also isolated giant viruses from people, including a patient with pneumonia. It's not clear yet what role giant viruses play in our health. They may be able to directly infect our own cells, or they may just lurk harmlessly in amoebae that invade our bodies.

The story of giant viruses drives home just how little of the virosphere we have explored. And it injects fresh life into a long-running debate: what exactly is a virus?

As soon as scientists began to understand the molecular makeup of viruses, they realized that they were fundamentally different from the familiar forms of cellular life. When Wendell Stanley produced crystals of tobacco mosaic virus in 1935, he unsettled neat distinctions between the living and the nonliving. As a crystal, his virus behaved like ice or diamonds. But when provided with a tobacco plant, it multiplied like any living thing.

Later, as the molecular biology of viruses came into crisper focus, many scientists decided that they were only lifelike, but not truly alive. All the viruses that scientists studied carried a few genes apiece, leaving a wide genetic gulf dividing them from bacteria. The few genes that viruses carried allowed them to execute the barest tasks required for making new viruses: to invade a cell and slip their genes into a cell's biochemical factories. Missing from viruses were all the genes for full-blown life. Scientists could find no instructions in a virus for making a ribosome, for example, the molecular factory that turns RNA into proteins. Nor did viruses have genes for the enzymes that break down food in order to grow. In other words, viruses appeared to lack much of the genetic information required to be truly alive.

In theory, though, a virus might be able to gain that information and become truly alive. After all, viruses are not fixed in stone. A mutation might accidentally duplicate some of their genes, creating new copies that could later take on new functions. Or a virus might accidentally take up genes from another virus, or even from a host cell. Its genome could expand until it could feed, grow, and divide on its own.

While it was conceivable that viruses could evolve their way to life, scientists saw a giant wall in their way. Organisms with big genomes need a way to copy them accurately. The odds of suffering a dangerous mutation increase as a genome gets larger. We protect our giant genomes from this risk by producing error-correcting enzymes, as do other animals, plants, fungi, protozoans, and bacteria. Viruses, on the other hand, have no repair enzymes. As a result, they make copying errors at a tremendously higher rate than we do—in some cases, over a 1,000 times higher.

The high mutation rate of viruses may put a limit on their genome—and thus may prevent them from being truly alive. If a virus's genome gets too big, it is more likely to suffer a lethal mutation. Natural selection may therefore favor tiny genomes in viruses. If that's true, then viruses may be unable to make room for genes that would let them turn raw ingredients into new genes and proteins. They cannot grow. They cannot expel waste. They cannot defend against hot and cold. They cannot reproduce by splitting in two.

All those *nots* added up to one great, devastating *NOT*. Viruses were not alive.

"An organism is constituted of cells," the microbiologist Andre Lwoff declared in a lecture he gave when he accepted the Nobel Prize in 1967. Not being cells, viruses were considered little more than cast-off genetic material that happened to have the right chemistry to get replicated inside cells. In 2000, the International Committee on Taxonomy of Viruses made this judgment official. "Viruses are not living organisms," they flatly declared.

The committee was drawing a stark line between viruses and the living world. But within a few years, the discovery of giant viruses blurred the line. If a tiny genome is one of the hallmarks of a virus, then it's hard to see how giant viruses can be considered viruses at all. Scientists don't know what giant viruses do with all of their genes, but some suspect that they do some rather lifelike things with them. Some genes in giant viruses encode enzymes that can repair DNA. They may use these enzymes to fix damage they might incur while traveling from one host cell to another. Many giant viruses carry genes for enzymes that assemble proteins—the kind of task that scientists thought only cellular life forms could carry out. It's possible that giant viruses flood their host with these protein-building enzymes to steer their metabolism in a new direction—one that benefits the viruses.

And when giant viruses invade amoebae, they don't dissolve into a cloud of molecules. Instead, they set up a massive, intricate structure called a viral factory. The viral factory takes in raw ingredients through one portal, and then spits out new DNA and proteins through two others. Giant viruses may use their viral genes to carry out at least some of this biochemical work.

The giant virus's viral factory, in other words, looks and acts remarkably like a cell. It's so much like a cell, in fact, that La Scola and his colleagues discovered in 2008 that it can be infected by a virus of its own. This new kind of virus, which they named a virophage, slips into the virus factory and fools it into building virophages instead of giant viruses.

By 2019, scientists had found 10 different virophages. They thrived everywhere from Antarctic lakes to the guts of sheep, and it's likely that many more remain to be found. Virophages are not merely parasites of parasites. They help cellular life forms by killing the giant viruses that make them sick. Even if a host cell dies from a giant virus infection, virophages will ensure there are fewer viruses to kill other cells. Scientists have found that algae carrying virophages grow into bigger blooms, likely because they carry protection against giant viruses.

For virophages and cells, these studies suggest, the enemy of my enemy is my friend. Some host cells even permit virophages to use their own DNA to store their genes. The virophage genes only come to life when a giant virus infects their host. They assemble into new virophages to attack the invader. Another line blurred: is the virophage a virus of its own, or a weapon deployed by the host cell? The answer may be a false choice. The interests of the virophage and the host cell align: they both want to destroy giant viruses for their own benefit.

Drawing dividing lines through nature can be scientifically useful, but when it comes to understanding life itself, those lines can end up being artificial barriers. Rather than trying to figure out how viruses are not like other living things, it may be more useful to think about how viruses and other organisms form a continuum. We humans are an inextricable blend of mammal and virus. Remove our virus-derived genes, and we would die in the womb. It's also likely that we depend on our viral DNA to defend against infections. Some of the oxygen we breathe is produced through a mingling of viruses and bacteria in the oceans. That mixture is not a fixed combination, but an ever-changing flux. The oceans are a living matrix of genes, shuttling among hosts and viruses.

While it's clear that giant viruses bridge the gap between ordinary viruses and cellular life, it's not yet clear how they reached that ambiguous position. Some researchers argue that they started out as ordinary viruses and then stole their extra genes from their hosts. Others have proposed instead that giant viruses existed at the very dawn of cellular life and evolved into more virus-like forms.

Drawing a bright line between life and nonlife doesn't just make it harder to understand viruses. It also makes it harder to appreciate how life began. Scientists are still trying to work out the origin of life, but one thing is clear: it did not start suddenly with the flick of a great cosmic power switch. It's likely that life emerged gradually, as raw ingredients like sugar and phosphate combined in increasingly complex reactions on the early Earth. It's possible, for example, that single-stranded molecules of RNA gradually grew and acquired the ability to make copies of themselves. Trying to find a moment in time when such RNA-life abruptly became "alive" just distracts us from the gradual transition to life as we know it.

In the RNA world, life may have consisted of little more than fleeting coalitions of genes, which sometimes thrived and sometimes were undermined by genes that acted like parasites. Some of those primordial parasites may have evolved into the first viruses, which may have continued replicating down to the present day. Patrick Forterre, a French virologist, has proposed that in the RNA world, viruses invented the double-stranded DNA molecule as a way to protect their genes from attack. Eventually their hosts took over their DNA, which then took over the world. Life as we know it, in other words, may have needed viruses to get its start.

At long last, we may be returning to the original two-sided sense of the word *virus*, which originally signified either a life-giving substance or a deadly venom. Viruses are indeed exquisitely deadly, but they have provided the world with some of its most important innovations. Creation and destruction join together once more.

ACKNOWLEDGMENTS

A Planet of Viruses was funded by the National Center for Research Resources at the National Institutes of Health through the Science Education Partnership Award (SEPA), grant no. R25 RR024267 (2007–2012), Judy Diamond, Moira Rankin, and Charles Wood, principal investigators. Its content is solely the responsibility of the author and does not necessarily represent the official views of the NCRR or the NIH. I thank the many people who advised this project: Anisa Angeletti, Peter Angeletti, Aaron Brault, Ruben Donis, Ann Downer-Hazell, David Dunigan, Cedric Feschotte, Angie Fox, Matt Frieman, Laurie Garrett, Edward Holmes, Akiko Iwasaki, Benjamin David Jee, Aris Katzourakis, Sabra Klein, Eugene Koonin, Jens Kuhn, Ian Lipkin, Ian Mackay, Grant McFadden, Nathan Meier, Pardis Sabeti, Matthew Sullivan, Abbie Smith, Gavin Smith, Philip W. Smith, Amy Spiegel, Paul Turner, David Uttal, James L. Van Etten, Kristin Watkins, Joshua Weitz, Willie Wilson, Nathan Wolfe , and Michael Worobey. I am particularly grateful to my SEPA program officer, L. Tony Beck, and to my editor at the University of Chicago Press, Christie Henry, for making this book possible.

SELECTED REFERENCES

A Contagious Living Fluid

Bos, L. 1999. Beijerinck's work on tobacco mosaic virus: Historical context and legacy. *Philosophical Transactions of the Royal Society B: Biological Sciences* 354:675.

Kay, L. E. 1986. W. M. Stanley's crystallization of the tobacco mosaic virus, 1930–1940. *Isis* 77:450–72.

Roossinck, M. J. 2016. *Virus: An illustrated guide to 101 incredible microbes*. Princeton, NJ: Princeton University Press.

Willner D., M. Furlan, M. Haynes, et al. 2009. Metagenomic analysis of respiratory tract DNA viral communities in cystic fibrosis and non-cystic fibrosis individuals. *PLoS ONE* 4 (10):e7370.

The Uncommon Cold

Bartlett, N., P. Wark, and D. Knight, eds. 2019. *Rhinovirus infections: Rethinking the impact on human health and disease*. London: Elsevier.

Hemilä, H., J. Haukka, M. Alho, J. Vahtera, and M. Kivimäki. 2020. Zinc acetate lozenges for the treatment of the common cold: A randomised controlled trial. *BMJ Open* 10(1).

Jacobs, S. E., D. M. Lamson, K. S. George, and T. J. Walsh. 2013. Human rhinoviruses. *Clinical Microbiology Reviews* 26:135–62.

Looking Down from the Stars

Barry, J. M. 2004. *The great influenza: The epic story of the deadliest plague in history*. New York: Viking.

Mena, I., M. I. Nelson, F. Quezada-Monroy, et al. 2016. Origins of the 2009 H1N1 influenza pandemic in swine in Mexico. *Elife* 5:e16777.

Neumann, G., and Y. Kawaoka, eds. 2020. *Influenza: The cutting edge*. Cold Spring Harbor, NY: Cold Spring Harbor Laboratory Press.

Taubenberger, J. K., J. C. Kash, and D. M. Morens. 2019. The 1918 influenza pandemic: 100 years of questions answered and unanswered. *Science Translational Medicine* 11:eaau5485.

Rabbits with Horns

Bravo, I. G., and M. Félez-Sánchez. 2015. Papillomaviruses: Viral evolution, cancer and evolutionary medicine. *Evolution, Medicine, and Public Health* 2015:32–51.

Chen, Z., R. DeSalle, M. Schiffman, et al. 2018. Niche adaptation and viral transmission of human papillomaviruses from archaic hominins to modern humans. *PLoS Pathogens* 14: e1007352.

Cohen, P. A., A. Jhingran, A. Oaknin, and L. Denny. 2019. Cervical cancer. *Lancet* 393:169–82.

Dilley, S., K. M. Miller, and W. K. Huh. 2020. Human papillomavirus vaccination: Ongoing challenges and future directions. *Gynecologic Oncology* 156:498–502.

Przybyszewska, J., A. Zlotogorski, and Y. Ramot. 2017. Reevaluation of epidermodysplasia verruciformis: Reconciling more than 90 years of debate. *Journal of the American Academy of Dermatology* 76:1161–75.

Weiss, R. A. 2016. Tumour-inducing viruses. *British Journal of Hospital Medicine* 77:565–68.

The Enemy of Our Enemy

Kortright, K. E., B. K. Chan, J. L. Koff, and P. E. Turner. 2019. Phage therapy: A renewed approach to combat antibiotic-resistant bacteria. *Cell Host & Microbe* 25:219–32.

Summers, W. 1999. *Felix d'Herelle and the origins of molecular biology*. New Haven, CT: Yale University Press.

The Infected Ocean

Breitbart, M., C. Bonnain, K. Malki, and N. A. Sawaya. 2018. Phage puppet masters of the marine microbial realm. *Nature Microbiology* 3:754–66.

Keen, E. C. 2015. A century of phage research: Bacteriophages and the shaping of modern biology. *Bioessays* 37:6–9.

Koonin, E. V., and N. Yutin. 2020. The crAss-like phage group: How metagenomics reshaped the human virome. *Trends in Microbiology*, February 28. https://doi.org/10.1016/j.tim.2020.01.010.

Koonin, E. V., V. V. Dolja, M. Krupovic, et al. 2020. Global organization and proposed megataxonomy of the virus world. *Microbiology and Molecular Biology Reviews* 84(2).

Zhang, Y. Z., Y. M. Chen, W. Wang, X. C. Qin, and E. C. Holmes. 2019. Expanding the RNA virosphere by unbiased metagenomics. *Annual Review of Virology* 6:119–39.

Our Inner Parasites

Chuong E. B. 2018. The placenta goes viral: Retroviruses control gene expression in pregnancy. *PLoS Biology* 16:e3000028.

Dewannieux, M., F. Harper, A. Richaud, et al. 2006. Identification of an infectious progenitor for the multiple-copy HERV-K human endogenous retroelements. *Genome Research* 16:1548–56.

Frank, J. A., and C. Feschotte. 2017. Co-option of endogenous viral sequences for host cell function. *Current Opinion in Virology* 25:81–89.

Hayward, A. 2017. Origin of the retroviruses: When, where, and how? *Current Opinion in Virology* 25:23–27.

Johnson, W. E. 2019. Origins and evolutionary consequences of ancient endogenous retroviruses. *Nature Reviews Microbiology* 17:355–70.

Weiss, R. A. 2006. The discovery of endogenous retroviruses. *Retrovirology* 3:67.

The Young Scourge

Bell, S. M., and T. Bedford. 2017. Modern-day SIV viral diversity generated by extensive recombination and cross-species transmission. *PLoS Pathogens* 13:e1006466.

Burton, D. R. 2019. Advancing an HIV vaccine; advancing vaccinology. *Nature Reviews Immunology* 19:77–78.

Faria, N. R., A. Rambaut, M. A. Suchard, et al. 2014. The early spread and epidemic ignition of HIV-1 in human populations. *Science* 346:56–61.

Gilbert, M. T. P., A. Rambaut, G. Wlasiuk, T. J. Spira, A. E. Pitchenik, and M. Worobey. 2007. The emergence of HIV/AIDS in the Americas and beyond. *Proceedings of the National Academy of Sciences* 104:18566.

Gryseels, S., T. D. Watts, J. M. K. Mpolesha, et al. 2020. A near full-length HIV-1 genome from 1966 recovered from formalin-fixed paraffin-embedded tissue. *Proceedings of the National Academy of Sciences* 117:12222–29.

Sauter, D., and F. Kirchhoff. 2019. Key viral adaptations preceding the AIDS pandemic. *Cell Host & Microbe* 25:27–38.

Becoming an American

Hadfield, J., A. F. Brito, D. M. Swetnam, et al. 2019. Twenty years of West Nile virus spread and evolution in the Americas visualized by Nextstrain. *PLoS Pathology* 15:e1008042.

Journal of Medical Entomology. 2019. Special Collection: Twenty Years of West Nile Virus in the United States. 56 (6). https://doi.org/10.1093/jme/tjz130.

Martin, M.-F., and S. Nisole. 2020. West Nile virus restriction in mosquito and human cells: A virus under confinement. *Vaccines* 8:256.

Paz, S. 2019. Effects of climate change on vector-borne diseases: An updated focus on West Nile virus in humans. *Emerging Topics in Life Sciences* 3:143–52.

Sharma, V., M. Sharma, D. Dhull, Y. Sharma, S. Kaushik, and S. Kaushik. 2020. Zika virus: An emerging challenge to public health worldwide. *Canadian Journal of Microbiology* 66:87–98.

Ulbert, S. 2019. West Nile virus vaccines—current situation and future directions. *Human Vaccines & Immunotherapeutics* 15:2337–42.

The Pandemic Age

Holmes, E. C., and A. Rambaut. 2004. Viral evolution and the emergence of SARS coronavirus. *Philosophical Transactions of the Royal Society B: Biological Sciences* 359:1059–65.

Morse, S.S. 1991. Emerging Viruses: Defining the Rules for Viral Traffic. *Perspectives in Biology and Medicine* 34:387–409.

New York Times. 2020. "He warned of coronavirus. Here's what he told us before he died." February 7. https://www.nytimes.com/2020/02/07/world/asia/Li-Wenliang-china-coronavirus.html.

Quammen, D. 2012. *Spillover: Animal infections and the next human pandemic.* New York: W. W. Norton.

Raj, V. S., A. D. Osterhaus, R. A. Fouchier, and B. L. Haagmans. 2014. MERS: Emergence of a novel human coronavirus. *Current Opinion in Virology* 5:58–62.

Tang, D., P. Comish, and R. Kang. 2020. The hallmarks of COVID-19 disease. *PLoS Pathogens* 16:e1008536.

Xiong, Y., and N. Gan. 2020. "This Chinese doctor tried to save lives, but was silenced. Now he has coronavirus." CNN. February 4, 2020. https://www.cnn.com/2020/02/03/asia/coronavirus-doctor-whistle-blower-intl-hnk.

The Long Goodbye

Duggan, A. T., M. F. Perdomo, D. Piombino-Mascali, et al. 2016. 17th century variola virus reveals the recent history of smallpox. *Current Biology* 26:3407–12.

Esparza, J., S. Lederman, A. Nitsche, and C. R. Damaso. 2020. Early smallpox vaccine manufacturing in the United States: Introduction of the "animal vaccine" in 1870, establishment of "vaccine farms," and the beginnings of the vaccine industry. *Vaccine* 38:4773–79.

Koplow, D. A. 2003. *Smallpox: The fight to eradicate a global scourge.* Berkeley: University of California Press.

Kupferschmidt, K. 2017. How Canadian researchers reconstituted an extinct poxvirus for $100,000 using mail-order DNA. *Science*, July 6. http://dx.doi.org/10.1126/science.aan7069.

Mariner, J. C., J. A. House, C. A. Mebus, et al. 2012. Rinderpest eradication: Appropriate technology and social innovations. *Science* 337:1309–12.

Meyer, H., R. Ehmann, and G. L. Smith. 2020. Smallpox in the post-eradication era. *Viruses* 12:138.

Noyce, R. S., S. Lederman, D. H. Evans. 2018. Construction of an infectious horsepox virus vaccine from chemically synthesized DNA fragments. *PLoS ONE* 13:e0188453.
Reardon, S. 2014. "Forgotten" NIH smallpox virus languishes on death row. *Nature* 514:544.
Thèves, C., E. Crubézy, and P. Biagini. 2016. History of smallpox and its spread in human populations. In *Paleomicrobiology of humans*. ed. M. Drancourt and D. Raoult, pp. 161–72. Washington, DC: ASM Press.
Wimmer, E. 2006. The test-tube synthesis of a chemical called poliovirus. *EMBO Reports* 7:S3–9.

The Alien in the Water Cooler

Berjón-Otero, M., A. Koslová, and M. G. Fischer. 2019. The dual lifestyle of genome-integrating virophages in protists. *Annals of the New York Academy of Sciences* 1447:97–109.
Colson, P., B. La Scola, A. Levasseur, G. Caetano-Anolles, and D. Raoult. 2017. Mimivirus: Leading the way in the discovery of giant viruses of amoebae. *Nature Reviews Microbiology* 15:243.
Colson, P., Y. Ominami, A. Hisada, B. La Scola, and D. Raoult. 2019. Giant mimiviruses escape many canonical criteria of the virus definition. *Clinical Microbiology and Infection* 25:147–54.
Oliveira, G., B. La Scola, and J. Abrahão. 2019. Giant virus vs amoeba: Fight for supremacy. *Virology Journal* 16:126.
Schulz, F., S. Roux, D. Paez-Espino, S. Jungbluth, et al. 2020. Giant virus diversity and host interactions through global metagenomics. *Nature* 578:432–36.
Zimmer, C. 2021. *Life's edge: The search for what it means to be alive.* New York: Dutton.

CREDITS

Chapter opening illustrations copyright © 2021 by Ian Schoenherr. Introduction: tobacco mosaic viruses, © Dennis Kunkel Microscopy, Inc. Chapter 1: rhinovirus, copyright © 2010 Photo Researchers, Inc. (all rights reserved). Chapter 2: influenza virus, by Frederick Murphy, from the PHIL, courtesy of the CDC. Chapter 3: human papillomavirus, copyright © 2010 Photo Researchers, Inc. (all rights reserved). Chapter 4: bacteriophages, courtesy of Graham Colm. Chapter 5: marine phage, courtesy of Willie Wilson. Chapter 6: avian leukosis virus, courtesy of Dr. Venugopal Nair and Dr. Pippa Hawes, Bioimaging group, Institute for Animal Health. Chapter 7: human immunodeficiency virus, by P. Goldsmith, E. L. Feorino, E. L. Palmer, and W. R. McManus, from the PHIL, courtesy of the CDC. Chapter 8: West Nile virus, by P. E. Rollin, from the PHIL, courtesy of the CDC. Chapter 9: COVID-19 image captured and color-enhanced at the NIAID Integrated Research Facility (IRF) in Fort Detrick, Maryland. Credit: NIAID (CC BY 2.0). Chapter 10: smallpox virus, by Frederick Murphy, from the PHIL, courtesy of the CDC. Epilogue: mimivirus, courtesy of Dr. Didier Raoult, Research Unit in Infectious and Tropical Emergent Diseases (URMITE).

INDEX